$19.00

CARPENTRY IN COMMERCIAL CONSTRUCTION

Byron W. Maguire

Craftsman Book Company
6058 Corte del Cedro, P.O. Box 6500, Carlsbad, CA 92008

Credits to Contributors

Many private institutions, corporations, and federal agencies have graciously permitted use of their work in this book. Without their assistance much of value to the reader would be lost. Their materials are excellent testimony to quality and depth of research that make the building of first-class structures possible.

The contributors and credits for their particular assistance are listed below:

The American Concrete Institute for data from their ACI Standards on forms construction and quality requirements (Chapter 2).

The American Plywood Association for data on plywood products for commercial application (Chapter 2, 3, 4, 5 and 6).

Armstrong World Industries for data on suspended ceilings (Chapter 6).

The Masonite Corporation for data on siding and paneling (Chapters 3, 5, 6 and 7).

The National Forest Products Association for tables and data on structural lumber (Chapters 2, 3, 4 and Appendix A).

The Southern Forest Products Association for Technical Bulletin No. 2 (Chapters 3, 4 and Appendix B).

The Superintendent of Documents for various articles and publications relative to commercial construction.

The United States Gypsum Company for data on drywall fire retarding and sound control and metal partition (Chapters 3 and 6).

The Western Wood Products Association for data on lumber and wood products (Chapters 2, 3, 4, 5 and 6).

LIBRARY OF CONGRESS
Library of Congress Cataloging-in-Publication Data

Maguire, Byron W., 1931-
    Carpentry in commercial construction / Byron W. Maguire.
        p. cm.
    Bibliography: p.
    Includes index.
    ISBN 0-934041-33-4 :
    1. Carpentry. 2. Building--Details. I. Title.
TH5606.M23 1988
694--dc19

88-1383
CIP

First edition ©1976 by Reston Publishing Company, Inc.
ISBN 0-87909-124-X
Second edition ©1988 by Craftsman Book Company

Contents

Overview of Commercial Carpentry

Commercial carpentry: the skilled methods used by carpenters working on structures classified as commercial and non-industrial; that is, townhouses, single-level office buildings, and the like

Coordinating: the act or acts in commercial construction of harmonizing the efforts of contractors, staff, subcontractors, and workmen who erect structures

Directing: in commercial construction, all the supervisory work by the contractor who heads the entire construction project

Planning: the accumulation and organization of all the production factors related to the building of a commercial structure by various people with specialized skills

Procuring: the planned hiring of men and purchasing of materials necessary for a commercial construction project

Residential carpentry: the skilled methods used by carpenters working on structures classified as residential; that is, single, detached houses

Scaffolds: the elevated platforms or planking that are used by workmen when their work area is above or beyond normal human reach

Structure: in this text, a townhouse, one-level office, store, or a restaurant building made from specially selected materials

Time-line plan: a management display chart on which the various phases of construction are plotted in relation to time and duration

Man is born curious and to live with adventure. His inquisitiveness leads him in search of both the unknown and the answers to his questions. A residential carpenter knows that there is commercial carpentry. He only has to look around to see it. But what does it hold that might provide challenge and lead him to success or failure, or awaken him to new learning?

The atmosphere is very different for the carpenter who works on commercial construction as compared to one who works on residential carpentry. A residential carpenter working on detached single houses often is involved in every phase of the construction from laying the footing boards, through shingling the roof, to completing the interior. Frequently he meets the new owners of the structure and has the opportunity to work with them. From this association he derives a genuine personal satisfaction for a job well done, even though he may be better at some phases of the work than others.

If, however, the carpenter is a member of a crew responsible for just one phase of residential construction, he becomes very adept at that one phase of work. This crew method is a mass production technique usually employed to build tracts of houses efficiently. The techniques of commercial carpentry are used. One crew does framing, another sheathing and door and window installations, another cornice and siding installations, and still another the roofing. Finally, a select crew completes the interior work. The carpenter in this latter type of work environment cannot easily obtain the same degree of satisfaction that comes from working on all phases of single-house construction. If he maintains or increases his level of skill, however, his peers and supervisors will quickly recognize that and will rely on his judgment and expertise. Usually he will then be promoted to a higher level of responsibility.

The work atmosphere in light commercial construction is somewhat like that of residential construction at which the owner is sometimes present and the architect frequently. The rapport that usually develops soon leads architect and carpenter to respect each other's skills and talents.

But the work atmosphere is also different. The carpenter on a commercial construction job may move from one crew to another as the building takes shape. He may build forms, then erect framing, and later do exterior and interior work. These tasks often require both the carpenter and his supervisor to broaden their skills. Let's begin to define the differences that the commerical carpenter will find in his new environment.

INTRODUCTION TO NEW WORK ENVIRONMENT

The scope of commercial construction for the carpenter varies considerably—from skyrise office buildings, through the shipbuilding and aircraft industries, to townhouses, one-level office buildings, and stores. The environment you will read about and study in this text and the principles you will apply are related particularly to the construction of townhouses, small office buildings, and stores.

From these designations you can see already that there are two distinct types of non-industrial commercial structures: office buildings and stores on the one hand, and townhouse residences on the other hand. Each has task requirements that differ significantly; however, there is also a significant overlap of carpentry tasks which apply to both types of structures. The study of townhouse construction provides a natural bridge of understanding from residential to light commercial construction.

Let's list simply but comprehensively the relationships and differences among the types of construction. See Table 1–1 to start with the overall picture:

TABLE 1–1:   RELATIONSHIPS BETWEEN TYPES OF CONSTRUCTION

Work Activities	Residential	Townhouse	Office Building
1. Formwork for concrete	5%	10%	40%
2. Framing	45%	50%	10%
3. Roofing	20%	15%	15%
4. Exteriors	15%	13%	10%
5. Interiors	10%	10%	18%
6. Cabinetmaking	5%	2%	7%

Beginning with the next chapter each work activity in Table 1–1 is examined thoroughly and logically chapter by chapter. The magnitude of the differences of work in different environments is specially discussed. But, to understand the general implications of the data in Table 1–1 consider these factors briefly.

Concrete formwork is more extensive for townhouse construction than for detached houses because of design differences and because of the size of the units and the number of units being built. In addition, gardens, curbing, and sidewalks are usually formed for townhouse developments as well as trash locations and so forth. Often a swimming pool or two is built at the same time. But office buildings require the most extensive concrete formwork. In fact, proportionately there is more formwork than any other single activity done for office buildings.

In residential construction, framing work consumes the majority of time on the job (45%). It consists of making floors, walls, partitions, sheathing and doing the numerous tasks associated with this work. Townhouses usually require slightly more framing (50%), while office buildings usually have limited framing requirements (10%).

Building the roof of a structure usually requires considerable time and effort. As you can see in Table 1–1, residential roofs account for a greater percentage of time than do those on townhouses and office buildings. Even through more and more truss roofing is now used in residential construction, a greater proportion continues to be "stick built." The truss roof, however, is employed for most townhouse and office building construction.

Exterior carpentry activities are often difficult to predict completely because a variety of exterior materials, such as stone or brick, are not applied by carpenters. But certain man-made products do require the same work skills as lumber products and are sometimes used on townhouses and office building exteriors as well as on residential exteriors.

Interior work includes installing partitions, wall paneling, trim, doors and ceilings. Generally these activities are the same for both residential and townhouse construction. A larger percentage of the carpenter's activities may be spent in office construction, however, because of the extensive use of prefinished materials.

Finally, there is more cabinet work for the average single residence than for the townhouse and still more for office buildings because requirements for custom cabinets are frequent. Almost all cabinetmaking in townhouses simply consists of installing pre-built kitchens and bathroom vanities.

We have noted the overall differences and similarities of residential and commercial work activities. It follows that there must also be a different approach to managing a commercial construction project. Whether you are a contractor, a subcontractor, a foreman, or a journeyman your responsibilities to or for management will be significantly different. Even if you are now only a trainee it is essential, nevertheless, that you know and understand some of the management problems so that you will be able to support your boss better. Table 1–2 presents some areas of management concern and their relative importance in each work environment by showing the distribution of responsibility.

Looking over Table 1–2 will help you appreciate that contractors often use the skills of specialists on commercial jobs. The management work frequently subdivided, includes materiel procurement, training, and, often, the scheduling.

Let's examine each activity briefly. *Planning* means developing a schedule of action so a job progresses in right order. Having men and material available at precisely the right moment is, of course, the ultimate objective of all job plans. This may be a complex task, espe-

TABLE 1–2: RELATIONSHIPS IN MANAGING ACTIVITIES

Organizational Function	Residential Construction	Townhouse/Commercial Construction
Planning	Contractor	Contractor/Superintendent
Coordinating	Contractor/Architect	Contractor/Architect/Foreman
Training	Contractor/Journeyman	Field Foreman/Contractor
Procuring	Contractor	Supply Specialist/Contractor
Directing	Contractor/Foreman	Contractor/Superintendent

cially if activities are interrelated or if other contractors are involved. A good plan accounts for the major operations and takes into account the possible circumstances that could alter them. A *Time-Line* plan like that in Figure 1–1 visualizes all phases of the work and the amount of time allowed for each at various stages in the schedule.

Coordinating is the harmonious arranging of activity—between contractor and foreman, between contractor and architect, between one contractor and other contractors, and between foreman and architect. Coordinating also involves liaison or lateral trust and reliance on a man to man basis and on a crew to crew basis. In addition, coordinating means ensuring that materials are supplied on time to the job, and inspecting each phase of the work. Coordination also ensures that all people on the job remain in contact and are mutually advised and informed of either availability or delays so that work proceeds on schedule.

Training is a two-fold organizational activity. First, someone in management defines skills requirements and then translates them into skills on hand, skills needed, and skills to be trained for. Secondly, the same manager (or his designate) develops a training plan that assures that the skills of the appropriate workmen are available when needed. As shown in Figure 1–2, a single form can record both the needed job skills and the qualifications and limitations of available personnel.

Procuring is the obtaining of materials and men needed for each work phase. It is particularly important for successful management. Since the various job deadlines are critical to construction schedules, a great deal hinges on good procurement methods. The materials, machinery, and supplies necessary for job completion must be ordered so that deliveries either precede or coincide with on-site usage. Work-

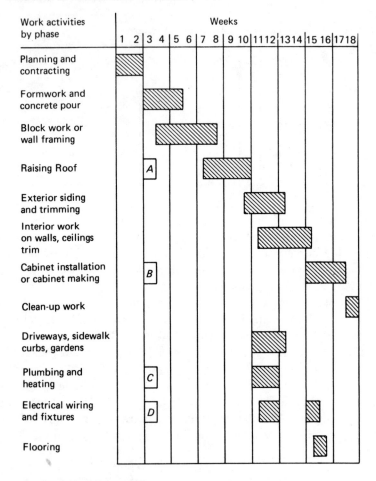

A Time for ordering trusses
B Time for ordering cabinets and display shelving
C Time to do rough plumbing
D Time to do underground electrical work

Figure 1–1 Time-Line Plan for Commerical Construction

men must be hired and/or trained according to job requirements, and, if specialized skills are needed, those responsible for procuring them must calculate sufficient lead time. This essential phase of procurement is usually interfaced with the training plan.

Directing in the commercial construction environment is also a two-fold activity. At the top management level the contractor directs the whole job, seeming sometimes much like an eight-armed octopus. Each organizational activity is also assigned a director. Sometimes, of

Name of Employee	Skill	Formwork	Framing	Roof Constr.	Ext. siding	Ext. trim	Int. walls	Int. ceilings	Int. trim	Cabinet work	Remarks

Note: The work areas shown above may be better defined by noting specific skills. Each chapter has a similar in-depth training record for each phase of work for each employee.

Skills code

☐ Basic ⬜ Skilled Ⓐ Apprentice

◼ Some Ⓢ Specialist ☐ None

Figure 1–2 Training Record of Construction Crews

course, the man in charge may wear more than one hat and function as director of several activities. Those who receive the closest direction are the crew members. They may not require individual direction on every task performance but they often are given new orders weekly, daily, and sometimes hourly.

From this somewhat lengthy introduction, you can begin to understand some of the differences between residential carpentry and commercial carpentry. But before you proceed with discussions of task and organizational activities, it is important to recognize that safety requirements are rules provided for your use and well-being.

SAFETY RULES

Safety on the job is very important. National agencies like the Department of Labor and the FHA concern themselves with the welfare of the worker on the job. Insurance companies are also concerned. Your own acceptance and implementation of safety practices will usually guarantee you a safe working environment.

Many factors must be included in any safety program. If they are carefully integrated in production plans several objectives can be accomplished in one operation. The following safety measures illustrate this point:

a. Plan for adequate storage of new, used, or reclaimed materials by selecting an area that ensures a clean working site. For example, as forms are stripped from walls and footings, they must be cleaned and then stacked out of the way until needed again. Consider—if a man has already lifted the board, how much more trouble is it to walk 30 feet with it and stack it out of his way?

b. Level and compact the ground during all excavation work. Grading and firming for a few minutes with the right machine can eliminate, even if only temporarily, any obstruction that might impede workmen. It will also provide a safe surface on which to build forms or erect scaffolds.

c. Build scaffolds properly and brace them with first-quality materials. Many men will walk, work, and stack materials on scaffolds so the proper reinforcement of supports, walking planks, and back rests is essential to good scaffolding.

d. Specify work clothes that promote personal safety: hard hats, steel-toed shoes with tie laces, shirts without trailing sleeves, etc. Remember, proper clothing contributes to job safety. Loose shoes and clothing are frequently the cause of on-site accidents; and the unfortunate thing is that most such accidents could have been avoided.

Safety of personnel cannot be overemphasized. Consider that a given work force is limited. If only the minimum of essential skills and experience are available for the job, the temporary loss of even one man through a serious accident may result in total loss of profit on the job and, even worse, in a personal loss to the workman and his family.

You will find out how important safety precautions and general preparedness are for qualified workmen engaged in commercial carpentry as you study the techniques of commercial formwork in the next chapter.

REVIEW QUESTIONS

1. Define two aspects of commercial carpentry that differ from residential carpentry.
2. It is customary to find crews of carpenters specializing in separate phases of commercial carpentry. True or False?
3. What is a reason for the greater amount of concrete formwork done for office buildings?
4. Name the five major elements of an organization.
5. Explain the importance of coordination in a commercial construction job.
6. Who does the procuring for the organization and what are some of his responsibilities?
7. List four major areas of a job in which safety plays an important part.

2

Formwork

Batter board: a temporary structure made from 2 × 4 and 1 × 6 lumber and installed at the corners of building layouts; used to fasten foundation lines

Column: a reinforced concrete upright; often a part of a poured wall

Concrete: a mixture of cement, sand, gravel, and water in specific proportions which when hard and compact forms a solid mass; used as a building material

Footing: a reinforced concrete base or foundation upon which building walls are constructed

Form: usually a wooden structure installed in a specified manner so that it will contain concrete in liquid form until it hardens

Girder: a reinforced concrete beam, usually horizontal, which serves as a structure for the floor joists or the framework of a structure

Header: a structural member used to span an opening and support a load; also used with girder formwork

Key: a trapezoidal-shaped member imbedded in a footing form so as to create a depression for secure wall anchoring

Lateral pressure: the force and thrust exerted by liquid concrete against the sides of a form

"Man-week": a normal work week for a carpenter, as 35 hrs or 40 hrs

Mudsill: a treated timber of 2″ nominal stock; lowest sill in a structure's foundation; used as a ground stabilizer

Plastic concrete: a concrete mix at the state in which it is readily molded; changes its shape slowly if the mold is removed

Shore: a vertical or inclined falsework support; a prop, as a beam, placed under or against something for stability

Spreader: a metal or wood member used to maintain proper and accurate separation between form walls; also a piece of lumber used to keep a door's side jambs apart evenly during installation (Chapt. 5)

Strongback: a vertical timber sometimes used outside the wales as a part of the form bracing

Tie: a wire or rod used to maintain and keep separate the walls of formwork under stress from the weight of concrete

Wale: a horizontal member of a form support; frequently used in pairs and held in place by spreader tie fasteners or strongbacks.

It is genuinely important that you have a total understanding of each phase of a commercial construction job as well as the whole job. Some data is very much like that used in residential carpentry, other data is unique to commercial applications. The organizational functions proper to formwork are discussed first. Following this, details and descriptions of job activities are considered; and finally, the training elements are defined. This last section is divided so the training elements consist of two items. One, the training chart, depicts the skills and knowledge needed for formwork; the other groups related Task Activities. Each task is developed so as to serve both as a learning and a measuring tool.

ORGANIZATIONAL FUNCTIONS FOR FORMWORK

Organizational functions are best visualized by the flow chart or functional diagram method, as in Figure 2–1. It charts the numerous activities which must be performed within the organization that deals with formwork. Most of the numerous elements are self-explanatory.

Note that directing the formwork is shown here as the contractor's function. That would not be true on a large project. In that case the contractor assigns a vice-president or a superintendent to direct the operation. Subcontracts are usually issued and signed by the contractor. Since it is not in the subcontractor's direct organizational structure, this function is shown to the right on the chart. Clerical operations and scheduled conferences are also indirect activities and, therefore, are also shown to the right of planning.

Within the rectangles under production the empty squares should be filled with the figures estimated in man-weeks for each work activity. The term "man-week" may be new to you. It simply specifies the traditional workweek (35 or 40 hours in most crafts) that each man is expected to work. It does not define and has nothing to do with pay or skill, or overtime; however, these factors do ultimately influence the assigned man-week figure when production is calculated for a given schedule. Where applicable you may indicate "man-days" in lieu of man-weeks.

Under the materiel listing in the flow chart there is an entry for "long-lead items". Materials in this category usually require from a month to a year to obtain. Since production and scheduling are integrated with materiel procurement, these long-lead orders must be carefully timed and placed.

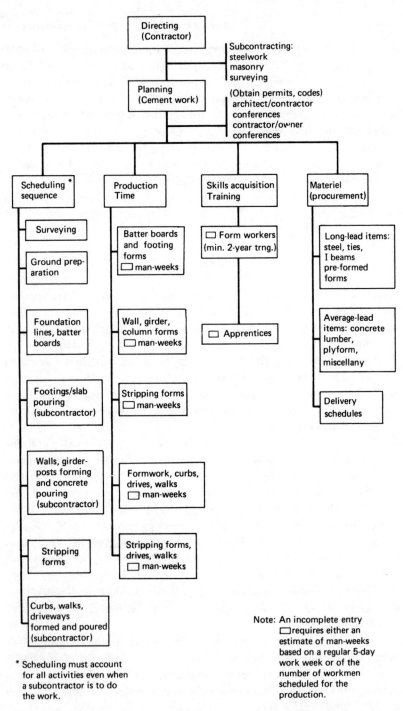

Figure 2–1 Organization Chart of Concrete Formwork Activities

Study and use this functional chart on formwork activities. Then either use it or make one adapted to your own organization. From its data develop a *Time-Line Plan* for this phase of the total job.

Note also the different task activities that are indicated. Let's study the details and descriptions in terms of the materials, methods, and procedures involved.

DETAILS AND DESCRIPTIONS OF FORMWORK ACTIVITIES

In a residential construction job you begin formwork by laying out the batter boards and foundation lines and continue with installation of footings and slabs. These tasks must be done in commercial construction; however, more extensive formwork is usually required. First, more substantial footings, curbings, sidewalks and driveways are needed. They are usually wider and deeper and so the specifications often call for different forming with greater support. Secondly, *walls, headers, girders* and *columns* requiring extensive formwork are the rule for commercial jobs rather than the exception. In this sort of forming you must learn about numerous elements and their functions, including *wales, ties, spreaders,* and *mudsills*.

Standards for Materials

Two concerns will usually govern the selection of materials for the forms: one, load-bearing limits; two, kind of re-use anticipated. If, on the one hand, the material is to be used afterwards for studding and sheathing, its quality, strength, dimensions, and overall utility must be selected with those functions in mind. If, on the other hand, the material is to be used repeatedly and exclusively for forming, then its working stress qualities need only to be established from that point of view. Forming materials range from stock lumber through plywood and fiberboard to plastics and metal. Generally, for constructing one-story office buildings, townhouse foundations, columns and girders or headers the forms will be designed and built from plywood, stock lumber, and/or metal.

Plywood: The types and grades of plywood designed specifically for formwork are listed in Table 2–1. Note that these plywood panels are uniformly standard in both thicknesses and sizes. They may, therefore, be used as sheathing afterwards. From the listed veneer grades and recommended structural uses you can determine which are most desirable for general forming, for wall forming, and for post forming.

TABLE 2-1: Plywood Grade-Use Guide for Concrete Forms *(Courtesy of American Plywood Association)*

USE THESE TERMS WHEN YOU SPECIFY PLYWOOD	DESCRIPTION	TYPICAL TRADEMARKS	VENEER GRADE		
			Faces	Inner Plies	Backs
APA B-B PLYFORM Class I & II**	Specifically manufactured for concrete forms. Many reuses. Smooth, solid surfaces. Mill-oiled unless otherwise specified.	APA PLYFORM B-B CLASS I EXTERIOR 000 PS 1-83	B	C	B
APA High Density Overlaid PLYFORM Class I & II**	Hard, semi-opaque resin-fiber overlay, heat-fused to panel faces. Smooth surface resists abrasion. Up to 200 reuses. Light oiling recommended between pours.	HDO PLYFORM I EXT-APA 000 PS 1-83	B	C-Plugged	B
APA STRUCTURAL I PLYFORM**	Especially designed for engineered applications. All Group 1 species. Stronger and stiffer than PLYFORM Class I and II. Recommended for high pressures where face grain is parallel to supports. Also available with High Density Overlay faces.	APA STRUCTURAL I PLYFORM B-B CLASS I EXTERIOR 000 PS 1-83	B	C or C-Plugged	B
Special Overlays, proprietary panels and Medium Density Overlaid plywood specifically designed for concrete forming.**	Produces a smooth uniform concrete surface. Generally mill treated with form release agent. Check with manufacturer for specifications, proper use, and surface treatment recommendations for greatest number of reuses.				
APA B-C EXT	Sanded panel often used for concrete forming where only one smooth, solid side is required.	APA B-C GROUP 1 EXTERIOR 000 PS 1-83	B	C	C

*Commonly available in 19/32", 5/8", 23/32" and 3/4" panel thicknesses (4' x 8' size).
**Check dealer for availability in your area.

In addition, the ability of a given plywood to withstand the *lateral* pressures of plastic concrete must be taken into account. Factors that influence this selection are the pour rate (number of cubic feet of wet concrete filled per hour in the form) and the outdoor temperature. Another factor you must consider is whether the plywood's face grain will be fastened parallel to or at right angles to the formwork supports. If parallel, a type such as B-B Plyform is usually best because it is waterproof and has a suitable exterior finish. As a rule properly braced ¾" plywood is strong enough to contain the normal 150 pcf (pounds per cubic foot) of pressure. Study the data in the table for additional specifications before selecting form materials.

Lumber: Consider stock lumber requirements from the same point of view as plywood when you select it to construct and brace the formwork in commercial construction. The elasticity, stiffness, and purity (absence from decay, large knots, inherent weakness) of stock lumber all add to or subtract from the total strength of the form. Grades number one or two of nominal one-inch and two-inch stock should always be used. Be sure also to select lengths that ensure minimal splicing of vertical members and the proper use of breaking-joint techniques on any horizontal members where spans of two or more studs exist between splices. Frequently mudsills will also have to be installed. Lumber for mudsills should be treated against decay to maximize their life cycle. If studs, joists, and other members used in the forming process are also to be used for the wall, floor, and roof construction be sure to choose lumber grades which ensure that their original structural quality meets contract specifications.

Ties: Wall forms are additionally reinforced against outward bulge and displacement by the use of either simple ties or combination ties and rods. Two types of simple wire ties are often used with wood *spreaders.* The wire is passed around studs and wales and through small holes bored in the sheathing panels of the form (Figure 2–2). Then the spreader is placed as close as possible to the studs and the tie is made taut either by inserting a wedge (upper detail, Figure 2–2) or by twisting it with a small toggle (lower detail). When the concrete reaches the level of the spreader during the pour, the spreader is knocked out and removed. The parts of the wire inside the form remain in the concrete; after the form is removed the outside sections are surplus and are cut off.

In commercial construction particularly, wire ties and wooden spreaders have been largely replaced by various manufactured devices which combine the function of tie and spreader. The one shown in

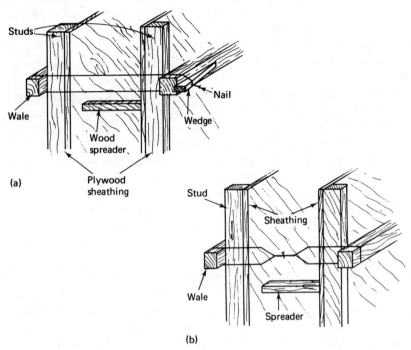

Studs

Wale

Wood
spreader

(a)

Plywood
sheathing

Nail

Wedge

Stud

Sheathing

Wale

Spreader

(b)

Figure 2–2 Types of Wire Ties

Figure 2–3 is called a snap tie. Such ties are made in various sizes to fit different wall thicknesses. The tie holders can be removed from the tie rod so the rod can be inserted in the small holes bored in the sheathing and pass between the horizontal wales. The wales are usually doubled for that purpose. The tie holders are tapped down on the ends of the rod in order to bring the sheathing to bear solidly against the spreader washers. To prevent the tie holder from coming loose, a double-headed nail is driven into the hole provided. Then, after the concrete has hardened, the tie holders are detached and the forms are stripped for re-use. After the forms are stripped, a special wrench is used to break off the outer sections of rod at pre-scribed points. These breaking points are located about an inch below the surface of the concrete. The small surface holes that remain can be plugged with grout as necessary.

Another type of wall-form tie much used in commercial work is the tie rod shown in Figure 2–4. It consists of an inner section threaded on both ends and two threaded outer sections. Cone nuts are threaded on the ends of the inner section and adjusted to the thickness of the

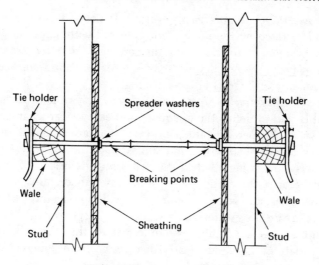

Figure 2–3 Types of Snap Ties

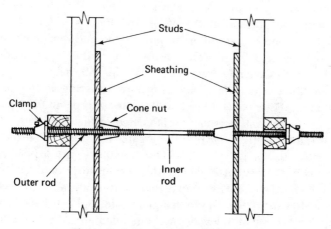

Figure 2–4 Detail of Tie Rod in Position

wall so that the base of the cone fits snugly against the inner form walls. The outer sections of the rod are passed through the wales and sheathing and threaded into the cone nuts in the form. Clamps are then threaded on the ends of the outer sections and tightened so that the forms bear uniformly against the cone nuts inside. After the concrete hardens, these clamps are loosened so the outer sections of rod can be unthreaded and removed from the cone nuts. After the forms are stripped, the cone nuts themselves are removed from the concrete by threading them off the inner sections of rod with a special wrench. The inner sections of rod remain in the concrete; the outer sections and the nuts may be re-used indefinitely. The cone-shaped surface holes which remain are plugged in the finishing.

Selecting Material to Meet Design Loads: There is a process that will enable you to define the adequate and safe design characteristics of formwork. Numerous tables provide standard calculations with which you can design, verify, and substantiate the soundness of your forms. This data is included in the following discussion of materials selection because the grade and type of materials dictate their placement, quantities, and related factors. You will see that the best quality lumbers can sustain greater loads while that of poorer quality means that significantly more lumber is required to do the same job. In other words, variations in design loads that result from different quality lumbers and related materials may mean cost differences of from several dollars to several hundred dollars.

Let's first study the various tables which identify specific pressure and stud-centering factors. Afterwards we can examine the method for determining the position of the wales and tie separations. Pressures for vibrated concrete at different pour rates and temperatures are shown in Table 2-2. These values are based on the recommendations of the American Concrete Institute and are for internal vibration only. These conditions are achieved by inserting a vibrating flexible metallic hose into the plastic concrete so that it causes settling and compacting. Among the important facts clearly indicated in the tabulated data we find that the faster the pour rate the greater the pressure on the form. For example, the pressure on the form is 410 psf with a pour rate of two feet per hour in 70°F; but if the pour rate is doubled to four feet per hour, the pressure at the same temperature is increased to 660 psf. Obviously then, both the pour rate and resulting pressure affect the selection of plywood and the placement of studs, wales, and ties.

Before proceeding, let's examine Figure 2-5 where the wall form is shown complete. Then we can relate the terms and the work as a whole. The wall form consists of plywood panels, studs, wales, and wall ties. Before the concrete is poured this form will be thoroughly braced for the lateral or side-to-side pressure. You will recall that the pour rate is the depth in feet at which the concrete is filled in the form per hour. The slower the pour rate the more time the concrete has to set or harden, thus the lower the pressure on the form.

TABLE 2-2: Concrete Pressures for Column and Wall Forms

Pour rate (ft./hr.)	Pressures (psf) of vibrated concrete[a,b,c]			
	50°F[d]		70°F[d]	
	Columns	Walls	Columns	Walls
1	330	330	280	280
2	510	510	410	410
3	690	690	540	540
4	870	870	660	660
5	1050	1050	790	790
6	1230	1230	920	920
7	1410	1410	1050	1050
8	1590	1470	1180	1090
9	1770	1520	1310	1130
10	1950	1580	1440	1170

Notes: a. Maximum pressure need not exceed 150h, where h is maximum height of pour:

b. For non-vibrated concrete, pressures may be reduced 10%.

c. Based on concrete with density of 150 pcf and 4-in. slump.

To return to the design process, let us again review Table 2–1 (page 14) in which plywood materials of Class I and Class II quality are identified. Table 2–3 shows the allowable pressures for plywoods with the face grain crosswise to formwork supports and Table 2–4 shows the allowable pressures for plywoods with face grain parallel to the formwork supports. Data in both these tables are for Class I type plyform panels. Similar tables available from The American Plywood Association identify values for Class II panels.

For example, using Table 2–2 and assuming a pour rate of 4 feet per hour, we find under the heading "70°F and walls" that the pressure is given as 660 psf. Next we refer to Table 2–3 under the heading "plywood thickness ¾ inch" and find that the closest value to 660 is 730. From this, we look left to the column "Support spacing" and read 12″ oc (for the studs). This means two things: the spacing of studs on 12″ centers provides a greater strength than required and it allows for a slight increase in the pour rate. Consider the cost factor for this form, however; it may not be essential to pour so quickly. Remember, a higher pour rate means more time and more money spent for bracing. If the pressure were reduced to 410 psf, you can see from Table 2–3 that the stud spacing could be 16″ oc. Now what is

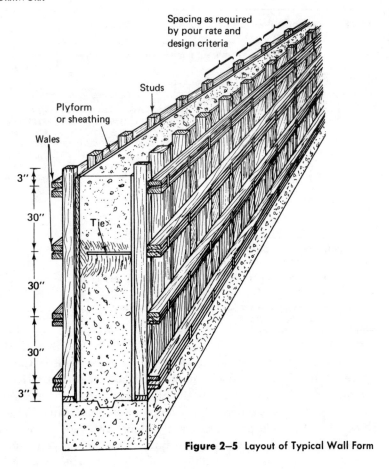

Spacing as required
by pour rate and
design criteria

Studs

Plyform
or sheathing

Wales

3″

30″

Tie

30″

30″

3″

Figure 2–5 Layout of Typical Wall Form

TABLE 2-3: Allowable Pressures for Plywoods with Face Grain Across Supports

Support Spacing (in.)	Plywood Thickness (in.)													
	15/32		1/2		19/32		5/8		23/32		3/4		1-1/8	
4	2715	2715	2945	2945	3355	3355	3580	3580	4010	4010	4110	4110	5965	5965
8	885	885	970	970	1215	1215	1300	1300	1540	1540	1580	1580	2295	2295
12	335	395	405	430	540	540	575	575	695	695	730	730	1370	1370
16	150	200	175	230	245	305	265	325	345	390	370	410	740	770
20	—	115	100	135	145	190	160	210	210	270	225	285	485	535
24	—	—	—	—	—	100	—	110	110	145	120	160	275	340
32	—	—	—	—	—	—	—	—	—	—	—	—	130	170

Data: Courtesy American Plywood Association

TABLE 2-4: Allowable Pressures for Plywoods with Face Grain Parallel to Supports

Support Spacing (in.)	Plywood Thickness (in.)													
	15/32		1/2		19/32		5/8		23/32		3/4		1-1/8	
4	1385	1385	1565	1565	1620	1620	1770	1770	2170	2170	2325	2325	4815	4815
8	390	390	470	470	530	530	635	635	835	835	895	895	1850	1850
12	110	150	145	195	165	225	210	280	375	400	460	490	1145	1145
16	–	–	–	–	–	–	–	120	160	215	200	270	710	725
20	–	–	–	–	–	–	–	–	115	125	145	155	400	400
24	–	–	–	–	–	–	–	–	–	–	–	100	255	255

Data: Courtesy American Plywood Association

the pour rate? Referring to Table 2-2, we see that 410 indicates a 2-ft per hour pour rate. Accordingly, concrete for an eight-ft high wall could be poured in 4 hours.

Apply the same technique to use data in Table 2-4 which concerns plywoods whose face grain is parallel to form supports. In the original example above (pour rate of 4 ft per hour = 660 psf) the ¾" plywood requires studs 12" oc. Table 2-4 shows that ¾" plywood requires studs with 10" oc spacing for the same pour rate. For 16" oc supports, the pour rate is reduced to 2 hours for plywood *across* supports, and to just about 1 hour for plywood *parallel* with supports. Now that you have learned how to use these calculations you know several factors:

1. Position of plyform face grain—vertical or across
2. Spacing of studs on centers
3. Pour rate per square foot

By keeping these factors in mind and by using the data in succeeding tables and figures, you can select the quality of stock lumber needed and define its placement in forms. Remember that the loads carried by slab joists and by wall studs and wales are proportional to their spacings as well as to the maximum concrete pressure. In addition, you can learn how to estimate the quantities of stock lumber used in forms.

Two grades of lumber are suitable for the construction of forms. The first listed is the best; the second merely adequate:

1. Douglas Fir, Larch No. 2, or Southern Pine No. 2 (19% moisture content)

2. Hemlock, Fir No. 2

There are other tables containing data relative to each grade at the end of the chapter (Tables 2-5 and 2-6).

The design characteristics of forms are further defined by the data in the figures which carry excerpts from the tables concerned with Douglas Fir, Larch, or Southern Pine No. 2.

Recall that in the example on page 19 the original plan was to have a 4-ft per hour pour rate and assume that the desired finish texture of the concrete dictated that the plywood panels be installed parallel to the supports. That in turn established the spacing of the studs at 10″ oc using ¾″ plyform panels. (Verify in Table 2-4). From the calculations in Tables 2-5 and 2-6 you can determine the maximum span a stud or joist will have between wales or supports and the maximum separation of wall ties along a given wall.

Figure 2–6 is an extract from Table 2-5 (Douglas Fir, Larch No. 2, or Southern Pine No. 2 with 19% moisture.) There are three main sections to the table, and several subsections as well. The left column gives the *equivalent uniform load* in pounds per foot. The second column lists conditions for determining what given length of a 2 × 4 or 2 × 6 can support a given load. The third column provides data similar to that in column two but for greater spans.

To use the table you must first determine the uniform load. Then to the right under column 2 or 3 read the member's maximum length in inches. This data then gives the on-center separation of the wales on a wall form.

Using the example of 660 psf of pressure, calculate the maximum span of stud between wales. Recall that stud spacing equals the maximum support spacing for the plywood. (In the earlier example, studs are spaced 10″ oc.) The *load per stud* equals the concrete pressure multiplied by the stud spacing in feet.

$$660 \text{ psf} \times \frac{12}{12} = 660 \text{ lb per ft}$$

Example: studs 10″ oc

$$660 \text{ psf} \times \frac{10}{12} =$$

$$660 \text{ psf} \times 0.83 = 550 \text{ lbs per ft}$$

Assume 2 × 4 studs continuous over three or more supports (wales). This figures out as four sets of wales for an 8 foot wall, for instance. From the calculations in Figure 2-6, you can see that 660 lbs falls between 600 and 800; and at the right in column 3, under 2 × 4 for

Equivalent Uniform Load (lb/ft)	Continuous Over 2 or 3 Supports (1 or 2 Spans)							
	Nominal Size							
	2x4	2x6	2x8	2x10	2x12	4x4	4x6	4x8
400	33	52	68	86	104	50	79	100
600	25	39	50	64	77	44	66	88
800	21	32	42	53	64	38	58	76
1000	18	28	37	46	56	32	49	64
1200	17	26	33	42	51	28	43	56
1400	16	24	31	39	47	25	39	50
1600	15	22	29	36	44	23	35	46
1800	14	21	27	35	42	21	33	43
2000	13	20	26	33	40	20	31	40
2200	13	20	25	32	39	19	29	38
2400	13	19	25	31	37	18	28	36
2600	12	19	24	30	36	17	27	
2800	12	18	23	29				

Equivalent Uniform Load (lb/ft)	Continuous Over 4 or More Supports (3 or More Spans)							
	Nominal Size							
	2x4	2x6	2x8	2x10	2x12	4x4	4x6	4x8
400	39	58	76	97	118	60	88	116
600	29	45	59	75	91	49	72	95
800	24	37	48	61	74	43	62	82
1000	21	32	42	53	64	38	56	73
1200	19	29	38	48	58	33	51	66
1400	17	27	35	44	53	29	45	59
1600	16	25	32	41	49	27	41	54
1800	15	23	30	38	46	25	38	50
2000	15	22	29	37	44	23	35	46
2200	14	21	28	35	42	22	33	44
2400	15	21	27	34	41	21	32	41
2600	13	20	26	33	39	20	30	
2800	13	20	25	32				

Figure 2–6 Maximum Spans for Joists or Studs in Inches

600 lb per ft, a 29" length is indicated. Continuing, under 2 x 4 for 800 lb per ft a 24" length is indicated. Coming between these two values gives a 27.5" span for 660 lb per linear ft or 4 double wales spaced evenly, with top and bottom wales about 6 inches from the edge of form. (To estimate the values of a function between two known values is called interpolating.) Refer again to Figure 2-6 which diagrams this calculation.

With the number and position of the wales established, the next phase is to determine the spacing of the ties. Tables 2-5 and 2-6 also provide this data.

To determine the separation of wall ties, calculate the uniform load by using the *spacing between wales* instead of the spacing of studs. Therefore, in the problem following:

Equivalent Uniform Load (lb/ft)	Continuous Over 2 or 3 Supports (1 or 2 Spans)							
	Nominal Size							
	2x4	2x6	2x8	2x10	2x12	4x4	4x6	4x8
400	33	52	68	86	104	50	79	100
600	25	39	50	64	77	44	66	88
800	21	32	42	53	64	38	58	76
1000	18	28	37	46	56	32	49	64
1200	17	26	33	42	51	28	43	56
1400	16	24	31	39	47	25	39	50
1600	15	22	29	36	44	23	35	46
1800	14	21	27	35	42	21	33	43
2000	13	20	26	33	40	20	31	40
2200	13	20	25	32	39	19	29	38
2400	13	19	25	31	37	18	28	36
2600	12	19	24	30	36	17	27	35
2800	12	18	23	29	35	17	26	33

Equivalent Uniform Load (lb/ft)	Continuous Over 4 or More Supports (3 or More Spans)							
	Nominal Size							
	2x4	2x6	2x8	2x10	2x12	4x4	4x6	4x8
400	39	58	76	97	118	60	88	116
600	29	45	59	75	91	49	72	95
800	24	37	48	61	74	43	62	82
1000	21	32	42	53	64	38	56	73
1200	19	29	38	48	58	33	51	66
1400	17	27	35	44	53	29	45	59
1600	16	25	32	41	49	27	41	54
1800	15	23	30	38	46	25	38	50
2000	15	22	29	37	44	23	35	46
2200	14	21	28	35	42	22	33	44
2400	15	21	27	34	41	21	32	41
2600	13	20	26	33	39	20	30	39
2800	13	20	25	32	38	19	29	38

Figure 2–7 Maximum Spans for Double Wales in Inches

$$\text{Pressure (psf)} \times \frac{\text{spacing}}{12} = \text{equivalent uniform load}$$

$$660 \text{ psf} \times \frac{27.5}{12} =$$

$$660 \times 2.29 = 1512.5 \text{ lbs per ft}$$

In Figure 2-7 you can see that 1512.5/2 = 756.25 lb/ft falls between 600 and 800, and to the right in column 3, under 2 x 4 you find that the spacing of wall ties must be between 29" and 24" or, interpolated, the values indicate that the ties should be spaced 27" oc.

Finally the design strength of the *tie* can be learned from data in this table. That value significantly affects the size of the tie that you order.

$$\text{Uniform load} \times \frac{\text{oc spacing of tie}}{12} = \text{allowable load of tie}$$

$$1512.5 \times \frac{27.5}{12} = \text{approx. } 3466 \text{ lb} \\ \text{of allowable load}$$

Before going on to the materials needed for bracing let's recap the techniques just studied in terms of uses. What you have learned is a method for designing a wall form, floor form, or column form. Discussion centered mainly on wall forming since you will probably make them most often. Usually the set of blueprints you work from will specify the size and placement of walls, floors, columns or girders. It is up to you to design the appropriate forms. Remember also that some specifications may list minimum wall tie load and pour rate.

From the various tables in this text you can easily verify the adequacy of the planned form. Then you will be secure in your understanding of the importance of pour rate.

Finally, you should now be alert to the specifications and requirements which directly help in estimating material requirements. Knowing the stud placement, number of wales, type and size of plywood, and quantity of ties for the job makes it possible to do a fairly accurate estimate. To make it complete, be sure to allow for sufficient bracing material.

Bracing or Shoring must accomplish several functions simultaneously. It must aid in and maintain form alignment; it must prevent pressures, such as those from wind and plastic concrete, from moving the form. Bracing must be installed so as to prevent the form from rising and must be strong enough to accept a distributive load of not more than 100 pounds per square foot.

Bracing normally is installed against the wale to a ground support. The angle of the brace is usually planned to be less than 45 degrees for wall bracing. Figure 2–8 illustrates several techniques. Notice that in each case the angle is shown at less than 45 degrees. (A brace with an angle greater than 45 degrees would actually help make the form rise since pressure to the form would thus originate on the opposite side to the brace or from within the form.)

Shoring is a way to brace a wall by installing vertical members either under a girder form or under a slab form. These members as well as the braces must be sound and equal the grade quality of the studs, wales, and joists. These requirements for quality materials are

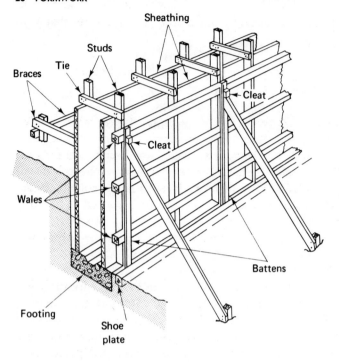

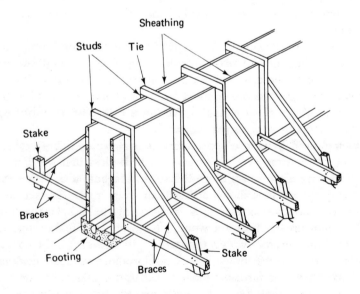

Figure 2–8 Layouts for Bracing Wall Forms

only part of the job, of course. The quality and standards of workmanship are also vital to the whole job.

Standards of Formwork

Whether the form is extensive, as for a wall, or simple, as for a sidewalk, you must maintain qualitative standards. To ensure that your formwork meets standards, you must know and understand its elements. The Quality Control Check List in this section will help since it includes the construction principles for good, safe, reliable formwork. The inspection list and related facts are excerpted from industry standards currently in force.

There is a general misconception about the quality of workmanship important to the construction of forms. Experienced carpenters realize that inexperienced carpenters often think that formwork doesn't require as much attention to standards of quality and performance as house trimming does. In fact, many inexperienced workmen seem to feel that any reasonably accurate form, even if barely able to contain the concrete, is sufficient for most jobs. Experienced carpenters, on the contrary, say that only after considerable training and understanding have they come to recognize the full extent of the quality workmanship necessary for good form making. Keep that in mind. Let the idea be your motivation for excelling at this extremely important phase of construction.

The American Concrete Institute has devised a standard (ACI 347-78) for construction of forms used in all variety of concrete structures. It incorporates concepts of form construction, including design considerations for load, strength, stresses, and tolerances. (See the technical reference list on page 273 for the address of the Institute if you wish to order a copy of the standard.)

You will need to use a blueprint as a construction guide while you build the form. Most of the form's structural elements will not be specified, however. It is your responsibility to ensure that the forms are soundly made. To this end you must maintain certain minimum building techniques and standards of workmanship. An inspection checklist such as the one below will quickly point up any shortcomings in the quality of your formwork. It will also help you in developing any needed training plan, since each item is a significant point that every trainee must understand and apply fully. Its use should result in inspections at which few construction faults can be found.

Typical Quality Control Checklist for Formwork

ITEM	SAT/N/A*	UNSAT**
1. Has each splice in a stud, wale or shore been properly made, secured and tied?	————	———
2. Has the staggering of joints or splices in sheathing, panels, wales, and braces been properly performed according to contract specs and accepted industry standards?	————	———
3. Has each shore and brace been properly seated and secured?	————	———
4. Has the proper number of form ties or clamps been located and installed according to data from specs and tables?	————	———
5. Is each tie or clamp properly tightened?	————	———
6. Before installation of mudsills was bearing soil inspected and corrected if found poor or inadequate?	————	———
7. Have all connections of shores, joists, wales, studs and braces been checked to ensure that form will not uplift or be wrenched by torsion at joints?	————	———
8. Have form coatings been applied *before* placement of steel, but not in such quantities as to run onto bars or footings?	————	———
9. Have expansion joints been installed and inspected?	————	———
10. Have all plywood joints been checked for alignment?	————	———
11. Does form alignment meet industry standards for accuracy within allowable variations? ($\frac{1}{4}''$ per 10 ft for plumb; $\frac{1}{4}''$ in 10 ft for level; in any bay of 20 ft, maximum $\frac{3}{8}''$; in forty feet or more, $\frac{3}{4}''$; in thickness of walls, slabs or beams $+ \frac{1}{4}''$ to $\frac{1}{2}''$; in footings $+ \frac{1}{2}''$ to $2''$)	————	———

*SAT = Satisfactory; N/A = Not Applicable
**UNSAT = Unsatisfactory

Inattention, neglect, omissions, cutbacks and short-cuts in labor and materials may result in the collapse of formwork and severe monetary loss. Industry estimates show that formwork and form materials account for 35 to 45 percent of the cost of the job. Let's now investigate some of this criteria for the items on the QC check list in order to clarify and expand on quality control.

Form Accuracy should be planned from the start. Precision instruments like the transit and level as well as mason line, plumb bob, and straightedge should be used all during construction phase and at each inspection of the work. Along with tools, bracing and shoring are important ways to keep the work in alignment.

No one tool is better, each has its purpose and application. For instance, the plumb bob is extremely easy to use to verify most perpendiculars. It is ideal for checking the vertical accuracy of walls and the position of footings in relation to foundation lines; however, it is difficult to use for plumbing a wall initially. A level and straightedge make a better combination for that task.

A transit instrument is a versatile instrument and it complements the mason line functionally. Both may be used initially, during construction, at QC checks, and during final alignment.

Form Support must always be adequate for the total design load. Earlier you read about the structural significance of correct measures for studs that span wales, the need to break joint on wales, the need for horizontal sheathing joints, and for numerous other elements important to formwork. Recall particularly that wales in need of splicing should have those splices off-centered by a minimum of two stud spans. Bracing, if spliced, should be reinforced with a "scab" or short member that nails on both sides of the butted ends. See Figure 2–9. Properly built forms sustain both the pour weight of wet concrete and the live pressure of plastic concrete; improperly built, they can bulge or collapse partially or totally.

The *mudsill* or footplate is the lowest frame member, usually 2 × 8 or wider and is used to provide support for the whole structure or just the bracing. It works on the principle of distributed mass. That is to say, its entire flat surface (the wide dimension) is laid flat against a ground surface; so the ground under the mudsill must be sound and uniform or the sill can not function properly.

To summarize, quality standards must be maintained in every aspect of formwork. That includes the selection of materials right for the job, the employment of trained labor to do the work, proper building techniques for the specific tasks, and finally a satisfactory QC program during the entire project. The last, quality control, is neces-

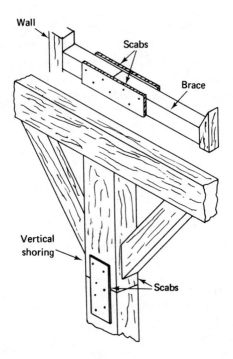

Figure 2-9 Layout for "Scabbing" a Joint

sary in order to monitor the accuracy, adequacy, and performance of all elements constantly and efficiently.

FORMWORK ACTIVITIES

Figure 2-10 illustrates common footings and forms for on-the-ground slabs. Note the similarities among the three types. The diagrams illustrate forms to a depth of 12 inches. Use nominal two-inch stock ordinarily because it is easier to brace, holds the plastic concrete more readily, and can be re-used more frequently than one-inch stock. Forms with a depth greater than 12 inches are usually made in panels similar to those shown in the detail at the right in the figure. Plywood or nominal one-inch stock is nailed to a 2×4 frame; then the frames are installed and braced to make the footing form. The procedures for making all these forms are essentially the same as for residential construction.

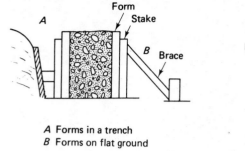

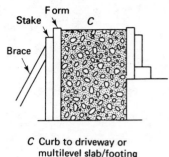

A Forms in a trench
B Forms on flat ground

C Curb to driveway or
multilevel slab/footing

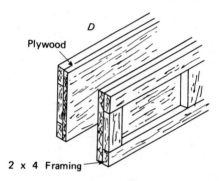

D Plywood panel W/2 x 4 framing

Figure 2–10 Footing Forms and Slab Forms

Foundation Preparations

The basic forming process is standard. Batter board assemblies are installed after the foundation corners or lines have been pinpointed by the survey team or the contractor. Ground is excavated, leveled, filled, and compacted according to building plan specifications and local code requirements applicable to the type of construction. Foundation lines that outline the foundation's perimeters are established on the batter boards (lower detail in Figure 2–11). Next the forms for footings and slabs are installed. Their exact positions are usually established, leveled, and made true with the plumb bob. The height of the formwork is dictated by plan requirements and the degree of level and accuracy stipulated by the institute (ACI) or contract "specs." Routines FW 1 and FW 2 (pages 44 and 45) give the procedural se-

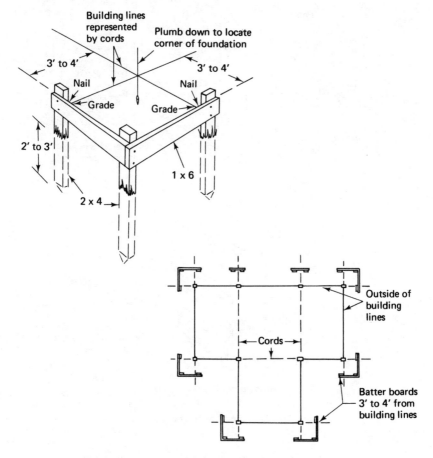

Figure 2–11 Batter Boards, Foundation, and Footing Lines

quences for tasks related to this work. By way of reminder, the routines can be used during actual construction, as a training vehicle, and as a quality control check list.

Study some of the numerous variations of these forms in Figure 2–12. Note that the graduated footing (A) has a piece of blocking at the end of the raised portion. Also note in Figure 2–12 (detail B) that forms for pilasters may either be incorporated in a standard footing form or built separately. If they are separately constructed, they must be soundly braced to prevent the form from lifting while the concrete pour is in progress.

Also note that if brick veneer is to be installed the footing can have more than one level. In Figure 2–12 (detail C) a 2×4 is installed

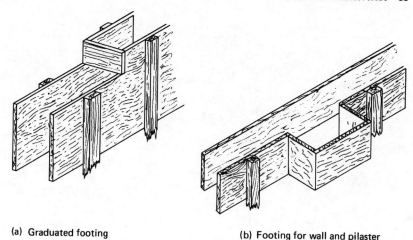

(a) Graduated footing   (b) Footing for wall and pilaster

(c) Offset in outside form
for brick veneer

Note: Removal of one side (usually inner)
makes a slab form

Figure 2–12 Variations of Footing Forms and Slab Forms

to create a depression in the footing so that the first course of brick can be laid below footing height and so below finished floor height.

Figure 2–13 illustrates the usual technique for creating curves in forms. Notice that a 2×2 stake is installed at the end of the straight run so that the material which makes the curve in the form can be nailed to it and *butted* to the straight run stock. Because such curved forms are generally made from thinner stock than the rest of the form, it is customary to space stakes closer than on straight runs and to add extra bracing as required.

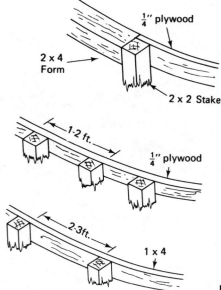

2 x 4
Form

$\frac{1}{4}''$ plywood

2 x 2 Stake

1-2 ft.

$\frac{1}{4}''$ plywood

2-3ft.

1 x 4

Figure 2–13 Wall Forms for Curves

Wall Forms

The several methods for constructing wall forms are essentially the same except that some use more structural material and bracing than others. You have already learned about the double wale and tie method of wall forming. A third type uses the *strongback* (Figure 2–14). A vertical member, the strongback is made a part of the form bracing and can be used whether single or double wales are used for the horizontal bracing.

The construction technique for on-site wall forms begins with the installation of a shoeplate on a previously poured and hardened footing. Sections of wall panels are then installed on this shoe. The wall panels usually consist of one sheet of plywood pre-drilled for ties and with three or four studs tacked to it. When a length of wall equal to the distance of a pour or wall length is in place, wales are tacknailed or toenailed where the ties come through the plywood or sheathing. The ties are then inserted and secured.

Insertion of the ties firms the wall and stiffens the cross points between studs and wales. The bracing, first used to hold up only a section of the form, is now installed at the proper angle to secure the wall as it is plumbed and aligned.

Before the opposite side of the wall form is installed the steel must

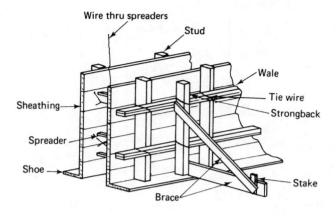

Figure 2–14 Wall Form Using Strongbacks

be installed. This task requires some time so planning is in order if the workmen's time is to be used properly. Stripping other forms and preparing more assemblies may be done.

All forms must be oiled or coated inside with a lubricant to prevent the concrete from adhering to the wood as it dries. The lubricant may be applied after one side of the form is up and just before the opposite side is raised.

Prepare the panels for the opposite form and raise it just as you did the first one. Secure the ties as soon as a wale is in position. This eliminates the need for several extra workmen and the excessive use of bracing.

Since you already have the first wall braced and accurately aligned, you must now brace the opposite side in a manner that does not disturb the wall alignment. Recall that all bracing must be installed so that neither the force of the wet concrete and the wind, nor any weight or stress from work tools, dolleys, or vibrators can cause the wall to lift or shift.

Exceptions in wall forms occur in almost every wall above ground and in many below ground. These "exceptions" are the openings for the passage of duct work and piping and for doors and windows. They have to be incorporated in the formwork and usually are built of stock lumber or plywood. They must be installed on the first wall or after both wall forms are erected. The thickness of the wall dictates the installation time and procedure. Figure 2–15 shows three types. One point which cannot be overemphasized is the need to reinforce these forms to prevent bending and shifting while concrete sets and hardens. Plastic concrete exerts its lateral pressure on all

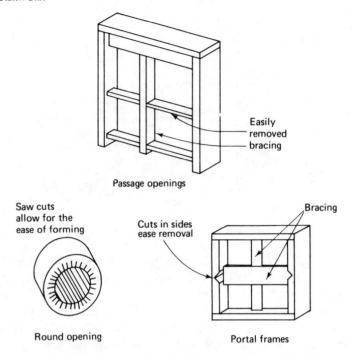

Figure 2–15 Wall Openings Used in Formwork

sides of the form at the same rate as it does on the outer wall forms.

Since the forms for the openings must be removed after the concrete is set and cured, careful consideration must be given to their construction initially. Figure 2–15 provides several clues.

Stripping of the forms should be done only after the concrete is cured properly. If care is exercised, there should be no loss of material, except for that used in the wall openings. Wall ties should be snapped, unscrewed, or clipped off in the concrete below the face of the wall. Then the tie holes can be plastered over by masons to finish off each wall area. Lumber and plywood salvaged from the formwork should be cleaned of nails and concrete and stacked or prepared for another section of the formwork.

Columns

A column is in effect a very small, tall wall section (Figure 2–16.) Its formwork is similar to that used for wall forming, except that it has to be made to take pressure on all four sides. In addition, since the pour rates tend to be faster than for long wall sections the psf is greater.

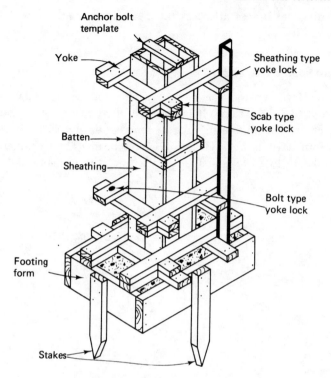

Figure 2–16 Form for Columns

The usual construction technique for a column form is to build four panels: two for the sides and two for the ends. Yokes are built around these; then bolts or ties are passed through the yokes and secured. Yokes stabilize the column form by functioning as wales do in wall forms. Their spacing is based upon the relationship of the lateral or sidewise dimension of the form to its height. Table 2–11 shows what that spacing or separation of yokes is for a given length. *For example,* to build a form with the right number of yokes for a column whose dimensions are 12″ wide by 24″ long by 8′2″ high, let us examine the data in Table 2–11.

Look in the stub or left-hand column until you find the column height (between 8′0″ and 9′0″); then move to the right until you locate the measurement under the 24″ column heading. That 16″ figure is the solution. The spacing or separation of yokes should be at every 16″ interval, with one at the base, one at the top and five between.

Follow standard methods of bracing for the column in position and for checking its alignment. Column forms have a greater tendency to lift than wall forms do, so be sure the base is well secured. Always install diagonal braces at a less than 45 degree angle from the base reference line.

Girders and Headers

So far as form-making specs go, a girder and header may be considered virtually the same. It is essential, therefore to build the forms strong enough to contain the plastic concrete until it sets. Figure 2–17 shows form layouts for a girder and for a beam. Their characteristics are similar to those for slabs and walls since the pressure of the con-

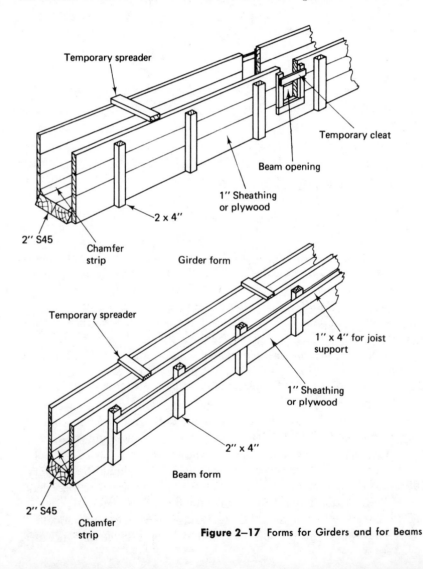

Figure 2–17 Forms for Girders and for Beams

crete is exerted both laterally and downwards (vertically). Studs and wales are used in the construction technique if the girder or header beam is large, or nominal two-inch stock if it is small. Standard procedures are followed and spreaders and ties can be used. To support the weight of a beam form, steel and concrete shoring must be installed underneath.

The length and girth of the girder will dictate the placement of the shoring and its size, e.g. 2 × 4, 4 × 4, 4 × 6. These dimensions will also influence the need for shoring "as you build" or after the form is already up. The shoring is toenailed to the girder base and actually sets either on a mudsill or on a floor or footing. At the upper end an angle brace of 1 × 4 or 2 × 4 should be installed to secure the shoring.

SUMMARY

To sum up: a variety of construction details and materials descriptions for all phases of formwork has now been examined. Task activities in formwork for footings, slabs, sidewalks, curbs, and driveways are similar to those performed in residential construction. But the forms for poured concrete walls, columns and girders present an interesting and new challenge.

The complex of structural materials used for them can be assembled with a minimum of fastening devices, but in such a way that the whole wall will be tremendously strong. Of course, proper bracing must be used also to prevent the forms from lifting, shifting, or collapsing. Remember this is best done by keeping the braces at a less than 45 degree angle to the ground level.

Even though the formwork activities were described in some detail, more in-depth descriptions are needed. This is especially so if you have never done this work before, or if your crews are inexperienced. The next part of this chapter provides the details for such training.

TRAINING IN FORMWORK

In this section the task activities are developed in procedural steps so as to present an in-depth understanding of formwork. Each step, clearly stated, accounts for a small unit of a total activity; and frequently several activities are used to complete a job. For example, you need to construct batter boards and set footing forms during the beginning phase of the job so erecting the boards and laying out their associated foundation line would be your first task activity. It is a stand-alone task. It has a beginning and an end. Installing footing forms is the second task activity. It too is a stand-alone with a definite beginning and ending. Scheduling for the two can be united, however, to complete a total phase of work since neither task alone would account for an entire job. From the numerous task activities in this section you can select some to do alone; you must group or sequence others. In this way every phase of the total job that requires their application will be handled efficiently.

As you study these task activities you will be learning invaluable work patterns: a) for the logical sequencing of certain tasks, b) for basing the assignment of those tasks upon the time-line plan for work phases, and c) for developing maximal manhour utilization through standardized procedures to assure production stability. Let's examine each of the three work patterns more closely.

a. The logical sequencing of tasks requires a listing of task activities in the order of their on-site performance and repetition:

1. Install footing forms for the whole structure
2. Install wall forms on 40-ft section of wall
3. Brace wall forms for first wall section
4. Remove footing forms and salvage material
5. Remove wall forms, salvage material, and reinstall on next 40-ft section
6. Install column forms for first wall section
7. Remove and re-install column form
8. Install curbing forms for whole project

The sample sequence above provides a logical succession of task activities about formwork since the tasks are listed as they are ordinarily done on the job. They also combine as a set or cluster of activities which complete a phase of the job. Notice that some are also repeated. This, of course, is necessary whenever conservation of materials is considered sufficiently important.

b. Time-line planning means charting the stages of multiple activities so that the workmen and the materials are on-hand at the right time in the right order. Each activity and its assignment must, therefore, be carefully dovetailed with the others. There are many aspects to consider in this assignment process: manhour availability, long lead-time items, allowances for delays, time lags in work during start-ups and for pre-preparation tasks, and the like. Remember all time blocks and lines of responsibility must be accounted for before a phase of construction can start. Example: curbing forms need to be removed before sidewalk forms are laid. Girder forms must be stripped before joists are installed. Planning for the performance of these tasks is more detailed and much more significant than just simple lists.

c. Maximal use of manhours by employing standardized procedures means that a given work effort is reduced to its basic elements. Each workman or group of workmen then learn to perform the same function repeatedly with a maximum of ease and a minimum of waste. In this way maximum production is achieved through motion economy. Modern assembly plants utilize this technique to the fullest extent. Fortunately, in carpentry many work actions do not repeat every five minutes or every 30 minutes or even every day, so the tedious side of standardization is largely eliminated. If the standardized approach to task completion is studied, understood, and performed skillfully by each member of the work force, both motion economy and maximum production will result. It is an approach that also reduces the need for continuous supervision.

With the knowledge that such a standard approach yields maximum production, you can make predictions about task time with a fair degree of accuracy. Usually you can substantiate any judgments on overall performance and you can use incentive programs to motivate workmen to above-standard performance. But more significantly, you can make overall projections on project scheduling and determine whether on-time or ahead-of-time completion is likely. In other words, the man-day or man-weeks in the time-line plan take on precise meaning.

The next consideration is how to go about implementing training. First, examine the carpenters and apprentices on the project. Find out about their theoretical knowledge and their practical experience in formwork. If they are weak in theory have them study material such as this chapter and that listed in the references at the back of the book. Make it a part of their training. Having them complete the quiz at the end of the chapter or talking over the theory behind the tasks are techniques that will identify their grasp of the work. Next, study their skill at practical application on the job for brief periods before the actual work in question begins. This promotes a standardized approach to task completion and gives training as well. For example, one or several task activities in this section can be used this way. You or the lead carpenter or the foreman should then preside over a training meeting. Read and discuss the task steps and correlate them with the building plans. That and the actual OJT surroundings will help to develop a unified approach to task accomplishment.

Recording the Training

Each analysis of work skills and knowledge and any special training sessions should be written down. Be sure to maintain individual records for and about the crews. They are, of course, required for all apprentice programs, manpower training programs and VA apprentice programs. It is only fair that a workman be able to learn how well he is doing and why and if he needs more training. For the most part, the objective data on the forms must express quantitative (time) values; however, excellence ratings such as superior, excellent, average, poor, or no training mean far more even though they are inevitably somewhat subjective. To simplify the "statusing" of a work force's abilities set up a training chart. It can be made quickly and easily. Figure 2–18 shows how one would look if filled out for formwork activities. Notice that personnel are listed at the left and the task activities across the top. Numerous methods can be used to record skill level, theory understanding, and degree of training. Make a key such as the one shown at the foot of the figure with a specific skill for each symbol. As each workman completes a phase of training or gives evidence of qualifications you can enter the appropriate symbol in the box cross-aligning with his name and under the appropriate task activity.

Task Activities

The next several pages contain the formwork task activity procedures. Each is well defined and is a tool to use for understanding, preplanning and on-the-job training (OJT).

Name of Employee	Skill	Footing forms	Batter boards	Foundation layout	Sidewalk and drive forms	Wall forms	Girder forms				Remarks

☐ None   ◨ Some   S Specialist

▱ Basic   ⊠ Skilled   A Apprentice

Figure 2–18 Training Record of Formwork Crew

FW 1 TASK ACTIVITY: FOUNDATION LAYOUT

RESOURCES

Estimated Manhours for 2 men: 8 to 16 hrs

Materials:

Floor plans, foundation layout, elevation reference drawing
3—2 × 4's each 4- to 5-ft long (for corner batter board stakes)
2—1 × 6's each 4 ft per corner board
#6d or #8d common nails
mason line, (footage equal to perimeters of foundation)
2 × 2 position stakes 12" long

Tools:

#8 crosscut saw	sledge hammer (10-16 lb)
hand hatchet	24" spirit level
portable power saw	16-oz hammer
transit	plumb bob
100-ft measuring tape	line level (*optional*)

PROCEDURES

1. Prepare batter-board stakes and batter boards and sufficient position stakes for job. Point all stakes.

2. After reading plans, use transit and 2 × 2 stakes to pinpoint 2 corners of building; drive a stake at each corner. Install a batter-board assembly 18" to 24" behind the stakes. BE SURE TO POSITION 1 × 6's SO THEY ARE EQUAL TO A SPECIFIC REFERENCE HEIGHT, such as the finished floor height.

3. Measure the length of an adjacent wall; position transit over one corner stake; sight to the other and swing transit 90 degrees; elevate transit sight to zero-in to the far end of the wall. Drive a 2 × 2 stake and pinpoint position with a nail driven into the top of the stake. Install another batter-board assembly.

4. Repeat Step 3 for the fourth corner; then locate intermediate points using similar techniques.

5. Using either the transit or a plumb bob and mason lines, install the lines on the batter-board assemblies. Verify the accuracy of the rectangles by cross-measurement at corners; and true-up the lines over 2 × 2 stakes with the plumb bob.

FW 2 TASK ACTIVITY: FOOTING AND SLAB FORMING

RESOURCES

Estimated Manhours for 1 man: 2.5 hrs @ 16 lin. ft
.5 hrs per corner

Materials:

floor plan and specification for foundations
1 × 3 or 2 × 2 stakes, 18″ to 24″ long & 4 ft apart, plus corner
   stakes
form material
#8d common nails
#6d common nails
bracing material (1 × 4's, 2 × 4's)

Tools:

24″ level	6-ft folding ruler
2-lb sledge hammer	plumb bob and line
16-oz claw hammer	flat shovel
100-ft mason line	#8 crosscut saw

PROCEDURES

1. Prepare stakes, and bracing, and footing forms if panels are to be used. Using plan, calculate stake requirements and cut enough for the footing or slabbing to be poured. (NOTE: a slab form is one-half of a footing form).

2. Dig and level the earth below the foundation line string on the batter board, as required by the plan. Lay the outside footing board on edge in the area. Suspend a plumb bob from the foundation line and measure from its point outward 4″ or as required by the plan. Butt the footing board against the ruler.

3. While holding the footing board erect, position a stake against its outside and at one end; and drive the stake into the ground. Repeat the procedure at the other end of the board and for any other footing boards necessary to complete the length of wall. Remove the plumb bob and use the line tied to the batter boards as a reference for establishing the height of the footings. (NOTE: a transit may also be used.) Position the forms for height and nail to stakes.

4. Repeat Steps 2 and 3 for all outside forms; then cut several spreaders equal to planned width of the footing (not required on slabs but useful on sidewalks).

5. Install the inside footing forms by placing one against two wooden spreaders; drive stakes on the outside of the forms; use a level to set the height of the form. Nail form to stakes when level.

6. Install additional stakes as required; brace the form with earth backfill and braces, as required; prepare keys if required.

7. Tack 1 × 2 straps across the tops of the form to hold keys in place.

8. Make curved forms from several thin pieces of material; but install in the same manner as regular form board. (NOTE: stakes may need to be spaced closer together to provide additional strength for the form.)

FW 3 TASK ACTIVITY: CURBING AND SIDEWALKS

RESOURCES

Estimated Manhours for 1 man: 1 hr @ 16 lin. ft.

Materials:

1 × 4's or 2 × 4's for form	1 × 3's for braces*
2 × 6's for form	expansion material
¼" × 4" (or 6") for curved sections	#6d common nails
1 × 3's or 2 × 2's for stakes	

Tools:

100-ft mason line (min. length)	#8 crosscut saw
6-ft folding ruler	hand ax
50-ft steel tape	2-lb sledge hammer
24" level	transit (*optional*)

*One per stake or as required; each should be cut so as to be equal to a length three times the form depth. For example, make 12" braces to brace a form 4" deep.

PROCEDURES

1. Prepare area for sidewalks and curbing by grading; establish grade levels with transit. Prepare stakes by cutting to length and tapering one end.

2. Install a form member along previously defined line; drive stakes behind form; raise form to proper grade. Nail stake to form at one end; repeat process at other end of the member. Drive additional stakes behind the form; nail to form. Brace form with backfill and additional wood braces.

3. Cut 2 spreaders from 1 × 3's to equal width of curb or sidewalk; position perpendicular to installed form. Position opposite form against spreaders and stake behind form. Using level and/or straightedge, raise form to level; nail to stake.

4. Repeat Steps 2 and 3 until all forms are installed.

5. Tie all butt joints made in forms by toenailing or backing 1 × 3's to them. Install expansion material as required.

6. Remove forms after curing process is complete.

FW 4 TASK ACTIVITY: WALL FORM

RESOURCES

Estimated Manhours for 2 men: 2 hrs per lin. ft for 8-ft high walls (avg.)

Materials:

shoe or soleplate timbers (2 × 4's, 2 × 6's, or 2 × 8's)
studs 2 × 4's
wales 2 × 4's or 2 × 6's
2 × 4's for braces
2 × 4's for stakes
2 × 10's (treated) for mudsill
ties @ type length and rating
#6d common nails
#8d common nails
#12d or #16d double-headed nails
4 × 8 sheets of ⅝" or ¾" plywood (plyform)

Tools:

16-oz hammer	transit
24" level	portable power saw
#8 crosscut handsaw	portable power drill
spade bits	2-lb sledge hammer
brace and bits	mason line
crow bar	50-ft steel tape
plumb bob	6-ft folding ruler
framing square	

PROCEDURES

1. Stack 3 to 4 sheets of plywood evenly. Lay out position for ties; drill required holes through all sheets. Repeat as needed.

2. For the form's first side, nail a 2 × 4 shoe outside at a distance from the wall line equal to the plywood's thickness.

3. Tack studs to the plywood's edge while plywood is on the stack. Raise plywood into position; brace temporarily. Toenail the studs to plate. Repeat the process until approximately 16 linear feet of wall are raised.

4. Install wales (single and/or double) along with snap ties or tie rods, starting from lowest wale. (Wire ties need wales nailed to studs.) Break joint on the wales by using various lengths of wale stock.

5. Nail studs to plywood, raise to position with ties protruding through plywood (for second side of the form, no shoe is required).

6. Tie walls together temporarily across the top with 1 × 4 scrap. Install wales from the ground up.

7. Align the wall. (This may have been done after the first side was installed.) With transit or mason line align the form's upper edge; plumb the wall simultaneously with straightedge and level or plumb bob; brace wall with 2 × 4 braces.

8. Miscellaneous tasks: oil forms; install necessary wall opening frames, such as window, door, pipes, duct openings; build scaffolds and runways.

9. Strip forms by removing scaffolds, bracing, wales, studs, and plywood, and finally the shoe. Clean stock of nails and cement; oil and stack for re-use.

FW 5 TASK ACTIVITY: GIRDERS AND HEADERS

RESOURCES

Estimated Manhours for 2 men: 4 hrs per 8 lin ft (avg.)

Materials:

nominal 2" base panels
side panels, or stock
ties, or wire for ties
shores (2 × 4s, 4 × 4's, 4 × 6's)
2 × 4's for braces
#6d common nails
#12d double-headed nails
chamfer strips
wales

Tools:

#8 crosscut handsaw
16-oz hammer
combination square
bevel square
level
mason line
portable power saw
portable power drill
spade bits
straightedge

PROCEDURES

1. Read plans to determine width, depth, and length of girder; lay out and cut base (soffit) stock, build side panels or cut stock for side panels and nail together.
2. Install and shore with posts properly braced; install chamfer strips and oil forms.
3. Strip forms after concrete is cured; remove all nails, stack lumber.

FW 6 TASK ACTIVITY: COLUMNS

RESOURCES

Estimated Manhours for 2 men: 8 hrs per 8 ft column (avg.)

Materials:

plyform sheets
studs
wales
tie rods
#6d nails per 8 ft column
#8d nails per 8 ft column

Tools:

portable power saw
#8 crosscut handsaw
16-oz hammer
24" level
framing square
ruler
crow bar
plumb bob

PROCEDURES

1. Obtain dimensions for the column from plans. Lay out and pre-pare four panels, one for each side; nail studs to panels; tack units together. Install chamfer strips if required.
2. Cut wales to form yokes around column panels and studs. If ties are used, fasten ties to wales. If external tie rods are used, fasten these through overhang on wales.
3. Position column form on footing; nail shoe to footing; install wales and ties and square form. Install braces; plumb column form.
4. Strip braces and wales after concrete has cured; then remove panels intact for re-use.

TABLE 2–5 51

TABLE 2-5: Maximum Spans For Joists or Studs Made of Douglas Fir-Larch No. 2 or Southern Pine No. 2 in Inches (Wales Separation)

Equivalent Uniform Load (lb/ft)	Continuous Over 2 or 3 Supports (1 or 2 Spans) Nominal Size								Continuous Over 4 or More Supports (3 or More Spans) Nominal Size							
	2x4	2x6	2x8	2x10	2x12	4x4	4x6	4x8	2x4	2x6	2x8	2x10	2x12	4x4	4x6	4x8
400	33	52	68	86	104	50	79	100	39	58	76	97	118	60	88	116
600	25	39	56	64	77	44	66	88	29	45	59	75	91	49	72	95
800	21	32	42	53	64	38	58	76	24	37	48	61	74	43	62	82
1000	18	28	37	46	56	32	49	64	21	32	42	53	64	38	56	73
1200	17	26	33	42	51	28	43	56	19	29	38	48	58	33	51	66
1400	16	24	31	39	47	25	39	50	17	27	35	44	53	29	45	59
1600	15	22	29	36	44	23	35	46	16	25	32	41	49	27	41	54
1800	14	21	27	35	42	21	33	43	15	23	30	38	46	25	38	50
2000	13	20	26	33	40	20	31	40	15	22	29	37	44	23	35	46
2200	13	20	25	32	39	19	29	38	14	21	28	35	42	22	33	44
2400	13	19	25	31	37	18	28	36	15	21	27	34	41	21	32	41
2600	12	19	24	30	36	17	27	35	13	20	26	33	39	20	30	39
2800	12	18	23	29	35	17	26	33	13	20	25	32	38	19	29	38
3000	12	18	23	29	35	16	25	32	13	19	25	31	37	18	28	36
3200	12	17	22	28	34	16	24	31	12	19	24	30	37	18	27	35
3400	11	17	22	28	33	15	23	30	12	18	24	30	36	17	26	34
3600	11	17	22	27	33	15	23	29	12	18	23	29	35	17	25	33
3800	11	17	21	27	32	15	22	29	12	18	23	29	35	16	25	32
4000	11	16	21	27	32	14	22	28	12	17	22	28	34	16	24	31
4500	11	16	21	26	31	14	21	27	11	17	22	27	33	15	23	29
5000	10	15	20	25	30	13	20	26	11	16	21	27	32	14	22	28

TABLE 2-6: Maximum Spans For Double Wales Made of Grade No. 2, in Inches (Ties Separation)

Equivalent Uniform Load (lb/ft)	Continuous Over 2 or 3 Supports (1 or 2 Spans)								Continuous Over 4 or More Supports (3 or More Spans)							
	Nominal Size								Nominal Size							
	2x4	2x6	2x8	2x10	2x12	4x4	4x6	4x8	2x4	2x6	2x8	2x10	2x12	4x4	4x6	4x8
400	28	44	57	72	87	48	73	96	33	51	67	85	103	54	79	104
600	22	33	43	55	66	39	59	78	25	38	50	64	77	44	64	85
800	18	28	36	46	56	32	49	64	21	32	42	53	64	37	56	73
1000	16	25	32	41	49	27	41	54	18	28	36	46	56	32	49	64
1200	15	23	30	37	45	24	37	48	17	25	33	42	50	28	43	56
1400	14	21	28	35	42	22	33	43	16	24	31	39	47	25	38	56
1600	13	20	26	33	40	20	31	40	15	22	29	36	43	23	35	46
1800	13	19	25	32	38	19	29	37	14	21	27	35	42	21	33	43
2000	12	19	24	30	37	18	27	35	13	20	26	33	40	20	31	40
2200	12	18	23	29	36	17	26	33	13	20	25	32	38	19	29	38
2400	12	18	23	29	35	16	25	32	13	19	25	31	37	18	28	36
2600	12	17	22	28	34	16	24	31	12	18	24	30	36	17	26	34
2800	11	17	22	27	33	15	23	30	12	18	23	29	35	17	25	33
3000	11	17	21	27	32	15	22	29	12	18	23	29	35	16	25	32
3200	11	16	21	27	32	14	22	28	12	17	22	28	34	16	24	31
3400	11	16	21	26	31	14	21	27	11	17	22	28	33	15	23	30
3600	11	16	20	26	31	14	21	27	11	17	22	27	33	15	23	29
3800	11	15	20	25	31	13	20	26	11	17	21	27	32	15	22	29
4000	10	15	20	25	30	13	20	26	11	16	21	27	32	14	22	28
4500	10	14	19	24	29	13	19	24	10	15	20	26	31	14	21	27
5000	9	13	18	23	28	12	18	24	10	15	19	24	30	13	20	26

*Spans are based on Ps-20 lumber sizes. Single member stresses were multiplied by a 1.25 duration-of-load factor for 7-day loads. Deflection limited to 1/360th of the span with $\frac{1}{4}$" maximum. Spans are center-to-center of the supports.

TABLE 2-7 53

TABLE 2-7: Recommended Maximum Pressures on Structural I Plyform (psf) (a) Face Grain Across Supports (b)

Support Spacing (in.)	Plywood Thickness (in.)												
	15/32		1/2		19/32		5/8		23/32		3/4		1-1/8
4	3560	3560	3925	3925	4560	4560	4860	4860	5005	5005	5070	5070	7240	7240
8	890	890	980	980	1225	1225	1310	1310	1590	1590	1680	1680	2785	2785
12	360	395	410	435	545	545	580	580	705	705	745	745	1540	1540
16	155	205	175	235	245	305	270	330	350	400	375	420	835	865
20	—	115	100	135	145	190	160	215	210	275	230	290	545	600
24	—	—	—	—	—	100	—	110	110	150	120	160	310	385
32	—	—	—	—	—	—	—	—	—	—	—	—	145	190

(a) Deflection limited to 1/360th of the span, 1/270th where shaded.
(b) Plywood continuous across two or more spans.

Data: Courtesy American Plywood Association

TABLE 2-8: Recommended Maximum Pressures on Structural I Plyform (psf) (a) Face Grain Parallel to Supports (b)

Support Spacing (In.)	Plywood Thickness (In.)												
	15/32		1/2		19/32		5/8		23/32		3/4		1-1/8	
4	1970	1970	2230	2230	2300	2300	2515	2515	3095	3095	3315	3315	6860	6860
8	470	530	605	645	640	720	800	865	1190	1190	1275	1275	2640	2640
12	130	175	175	230	195	260	250	330	440	545	545	675	1635	1635
16	–	–	–	–	–	110	105	140	190	255	240	315	850	995
20	–	–	–	–	–	–	–	100	135	170	170	210	555	555
24	–	–	–	–	–	–	–	–	–	–	–	115	340	355

(a) Deflection limited to 1/360th of the span, 1/270th where shaded.
(b) Plywood continuous across two or more spans.

Data: Courtesy American Plywood Association

TABLE 2–9   55

TABLE 2-9: Lateral Concrete Pressures For Various Temperatures

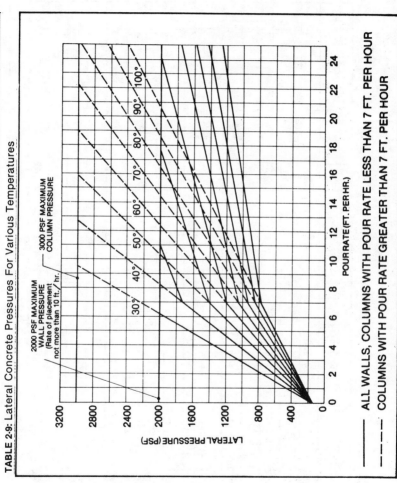

ALL WALLS, COLUMNS WITH POUR RATE LESS THAN 7 FT. PER HOUR

COLUMNS WITH POUR RATE GREATER THAN 7 FT. PER HOUR

Data: Courtesy American Plywood Association

TABLE 2-10: Specifications For Separation of Yokes on Column Forms

Largest dimension of column in inches = 'L'

Height	16″	18″	20″	24″	28″	30″	32″	36″
1′	31″	29″	27″	23″	21″	20″	19″	17″
2′								17″
3′				23″	21″	20″	19″	
4′	31″	28″	26″				18″	17″
5′				23″	20″	19″		
6′		28″	26″				17″	15″
7′	30″			22″	18″	18″	13″	12″
8′			24″		15″	18″	12″	11″
9′	29″	26″		16″	13″	12″	10″	10″
10′		20″	19″	14″	12″	12″	10″	8″
11′	21″		16″	13″	10″	10″	8″	8″ 7″
12′		18″		12″	9″ 9″	9″	8″	7″ 7″
13′	20″	16″	15″	11″	9″ 9″	8″	7″	6″ 6″
14′			14″	10″	8″ 8″	8″	7″	6″ 6″
15′	18″	15″	12″	9″	7″ 7″	7″	6″ 7″	
16′	15″	13″	11″	9″	7″ 7″	6″		
17′	14″	12″	11″	8″ 8″	6″ 8″			
18′	13″	12″	10″	8″				
19′	13″	11″	10″	8″				
20′	12″	11″	9″					

REVIEW QUESTIONS

1. Who has the director's responsibility in the management of form-work?

2. What is the definition of a *long-lead* item? Give an example.

3. Define the following terms: wale, wire tie, spreaders, snap tie, mudsill.

4. What type(s) of plywood is designed for forms?

5. How does the scheduled *pour rate* of liquid concrete affect the design of the wall form?

6. Is the *lateral* pressure of liquid concrete a side-to-side pressure?

7. How does the pour rate affect the on-center stud and wale spacing?

8. Given a pressure of 660 psf and spacing of wales at 30″ oc, what would the equivalent uniform load be?

9. List three functions of wall-form bracing and shoring.

10. What is the form accuracy standard that must be met for a wall form?

11. Explain the function of a *strongback*.

12. Name five forces that are controlled by proper bracing of a wall form.

13. What type of form makes use of a *yoke*?

14. What is the logical sequence of task assignment in forming activities?

3

Framing

Bridging: a method of reinforcing a joist foundation to make it more solid and free of movement; cross-bridging and solid bridging are the types most commonly installed

Dead load: a term used to account for the uniform, constant weight of all a building's materials and/or those applied to its floors and ceilings, including wallboard, plaster, subflooring, finished flooring, and the like

Deflection: the fractional percentage of inches that a joist bends when a load is placed upon it

Extreme fiber in bending: a numerical value given to a grade and species of lumber and based upon tests for fiber stress and rupture

Joist: a horizontal structural member laid on edge so as to rest on some points; holds up floors or ceilings; usually made from 2″ stock in widths ranging from 4″ to 12″ and in lengths up to and longer than 16′0″

Live load: the weight of inhabitants and furniture extensively applied to the floor of a structure; not constant in application, as walking people

Long-lead-items: the materials that require a three-month or longer period for delivery after orders are placed

Metal jamb: a metal jamb for a door; sometimes called a buck

Modulus of elasticity: a numerical value that represents a measure of stiffness in a structual member; the larger the number the stiffer the member

Plate: the top horizontal member of a wall framed with studs

Racking: the wrenching out of plumb or alignment of a wall, caused by external forces such as wind and lateral stress

Sill: the lowest installed structural member, made from 2″ stock; usually found on top of a cement foundation or wall; the member to which floor joists are nailed.

Soleplate: the lower horizontal member that supports a wall framed with studs

Toenailing: a method of nailing whereby the nail is driven slantingly and at an angle through the edge of one board into another's or through the side of a vertical plank to fasten it to the horizontal plank on which it is based

Framing is an essential part of any construction project. The primary objective of framing is, of course, the enclosure of designated space for certain functions. Many materials are available to do an effective job; and there are many techniques for construction, as well. This chapter examines the carpentry activities related to framing, except for roofing which is treated separately in Chapter 4.

Commercial framing techniques include all the task activities and materials used in residential construction and a few more. Girders, sills, floors, joists, partitions, sheathing, and specialized window and door units are among the familiar items covered in this chapter. Several newer approaches to framing are also identified. They include the 24-inch or MOD 24 construction method, the metal partition technique, and the soundproof floor-and-wall framing technique.

As in the previous chapter, you will first read and study the organization requirements that must be met in order to successfully complete this phase of commercial construction. Next, you will study the descriptions of and data on the numerous activities involved, as well as some of the special approaches, such as those used for MOD 24 work. Finally, you will study briefly a section on training. It is developed with a training chart and various tasks that are organized in procedural steps.

ORGANIZATION FUNCTIONS IN FRAMING CONSTRUCTION

All standard organizational activities are involved in this phase of commercial construction. Figure 3–1, illustrates rather well the degree of involvement for each activity. Let's review the procedural data in the figure before making any observations. Often a single conference between contractor and foreman may suffice to gather the particular information necessary for a mutual understanding of the framing tasks. Or, if the work is complicated, several conferences may be required.

All or some of the items listed under scheduling may apply to your job. You may know others that are not listed; if so you add them to the flow chart. Where a subphase is scheduled, it follows that you must have corresponding entries under "Production" and "Materiel." Remember, the scheduling of the numerous subphases of this phase will influence both the production and materiel procurement. Consider them carefully. Time and materials must also be allotted for all the work. You must also make sure skills are available on schedule. That means you have to examine the work records of all personnel thoroughly. Then if certain skills are lacking you must either plan for

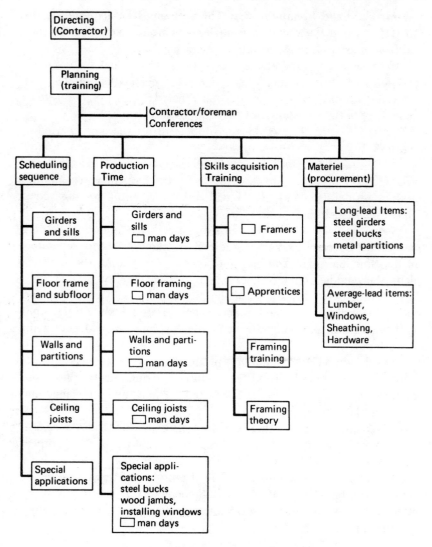

Figure 3–1 Organization Chart of Framing Activities

training or acquire the skilled workmen needed to handle such tasks.

Note carefully the provision for *man-days* under "Production" on the chart. Your own previous experience, time estimates from the Task Activities specs at the end of this chapter, and estimates from other sources will all help you to develop a useful reckoning of the number of man-days required for each subphase of work.

Under the entry "Apprentice Framers," two subjects are identified:

framing skills and *framing theory.* The skills must be acquired on the job (OJT) while the theory can be learned from books like this one and from master carpenters who are good teachers.

If you look carefully at the chart, you will note that several *long-lead-items* are listed under Materiel. Give special attention to these items and to others like them to ensure that deliveries of these items coincide with the scheduling of that subphase of work.

There are several general lessons to learn from the data charted in Figure 3–1.

First, except for work required with steel products most framing task activities are similar to those in residential carpentry. Techniques used in residential construction will, therefore, apply to most commercial construction as well.

Secondly, one or more of the subphases listed under scheduling may either be combined in a single time frame or listed separately on the time-line plan. You can find answers by reading the building plans thoroughly and by checking on the availability of materials. The contractor or his staff usually provide the delivery dates and delay data which guide the foreman. Appropriate scheduling decisions in this area are then generally a matter for foreman and contractor to work out.

Third, although scheduling, production, skills acquisition and materiel are separately listed, each with its requirements, these four planning sub-sections must be integrated for maximum efficiency and successful job completion.

In summary, examination of the construction plans and specifications, an estimate of materials, and the methodical use of this planning chart should enable you to identify and understand the problems, options, appropriate solutions, and final outcome of this phase of the work with a fair degree of accuracy. Remember that framing must be fitted into your time-line-plan since the work will, of course, influence all other phases of the job.

DETAILS AND DESCRIPTIONS OF FRAMING TECHNIQUES

The study of framing begins with the identification of various materials used in the work and then considers those methods and procedures that are applicable to commercial construction. Recent innovative developments include the 24-inch framing system and the steel-stud wall. Because of the proximity of families in townhouses and because of the heavy work flow in offices, soundproofing of rooms is also important. So the use of the framing techniques and insulation materials that further effective sound control will be examined as well.

Structural Characteristics of Lumber

There is very little difference between the lumbers you select for framing a residential structure and those you used in a commercial structure. Common members such as 2×4's and 2×6's, through 2×12's as well as 3″ and 4″ stock are used in both types of buildings. Construction plans give the nominal sizes and the specifications detail the grades. So in this section of the chapter we will try to understand why certain materials are chosen. We will also learn about the structural qualities of various sheathing grades and the different kinds of materials used for sheathing. This new data is significant for two reasons. First, commercial buildings are usually larger and accommodate more people than the average house so the selection of materials must be examined from the viewpoint of public use. Secondly, some understanding of the engineering principles on which the materials' grade and kind are based will help you to evaluate and accept the need for faithful adherence to the construction plan.

Among the new terms introduced in this study are *loads, deflection,* and *allowable stress.* Each identifies values applicable to structural joists used in framing, as well as roof rafters and general members. But this chapter concentrates on the joist in relation to framing problems. In the next chapter on roofs the terms will be applied to rafters.

Load: On either a floor or ceiling this means the total weight per square foot that the unit is predicted to support. This includes both the live and the dead load. The live load consists of the total furniture and fixture weight plus the total expected population weight divided by the square-foot area of the room or building concerned. The dead load consists of the weights of any products fixed to the joists, such as flooring, drywall ceiling materials, and insulation. In the average residential structure the standard loads are established at an average 40 psf (pounds per square foot) of live load and 10 psf dead load for activity rooms. Activity rooms include kitchens, dining rooms, living rooms, and the like. Sleeping or quiet rooms are usually predicted to have 30 psf live loads and 10 psf dead loads.

Translating this data to the townhouse, it is predictable that virtually the same criteria apply. Similar load allowances would generally fit specs for a small real estate or insurance office building. But a community center with a larger population capacity would require considerably more support because its predicted occupancy and use load could run 80, 90, or more pounds per square foot. So, with "load" defined adequately, the term can be used as a basis for selecting the proper member. There are several other terms that also need to be defined, however.

Deflection: For a structural member this means the degree to which it may safely deflect or bend under load. For appearance sake the deflection must also be limited so that sag will not be visible. In joist construction, the deflection factor may often be the controlling one in the determination of the size, grade, and species of lumber to be used. For structural joists, it is usually limited to one of three proportions:

$\int/360$ $\int$ = variable of span stated in inches

$\int/240$

$\int/180$

To provide the high level of stiffness needed in floor joists, for instance, specifications would be formulated for a 16-ft span thus:

$\int/360$ (span equals 16 feet converted to inches or $16 \times \dfrac{12}{1} = 192''$)

Accordingly we find that

$\int/360 = 192/360 = 0.533.$ or $\frac{1}{2}''$ deflection.

In contrast, using the other deflection formulas for the same 16-ft span results in these findings:

$\int/240 = 192/240 = .8$ or approximately $\frac{3}{4}''$ deflection

$\int/180 = 192/180 = 1.07$ or approximately $1\frac{1}{8}''$ deflection

Areas to which the deflection formula is applicable include the following:

$\int/360$ Floor joists in living and sleeping areas
          Ceiling joists with attic and plastered
          Ceiling joists with no attic if the live load is predicted 10 psf

$\int/240$ Ceiling joists with limited attic, drywall, and a live load of 20 psf
          Ceiling joists with no attic, drywall, and a live load of 10 psf

Notice that the $\int/240$ formula is not acceptable for floor joists and that $\int/180$ is not acceptable for any structural forming we are now considering. The $\int/180$ formula is applicable, however, to certain rafter calculations which we will use in Chapter 4.

Load and deflection characteristics are not the whole story. We must now study the meaning of allowable stresses and related data.

Allowable Stress: For a structural member it means that external force can be permitted to exert strain or pressure only so long as it does not permanently deform it or impair its function. Allowable stress is, therefore, concerned with two factors: Modulus of Elasticity, and Extreme Fiber in Bending. These are two values which express the ability of materials to resist and/or recover from various structural pressures.

Modulus of Elasticity: For a piece of lumber this is *the ratio between the deforming stress placed upon it and the corresponding fractional deformation of the fibers of the wood;* also called coefficient of elasticity. It is a measure of a material's ability to recover its original shape and size after being strained by a specific force. The Modulus of Elasticity symbol most often used in tabulated data is the letter "E". The numbers in columns listing the Modulus of Elasticity for various grades of lumber are usually expressed in millions of pounds per square inch (psi) or as fractional portions of millions of pounds per square inch. Table 3–1 summarizes values of Modulus of Elasticity so that allowable stress can be reckoned with in the selection process. The higher the numerical value assigned to a product the better the product resilience or ability to recover from stresses. Stated another way the greater the "E" value, the stiffer the product.

Determination of the allowable stress in lumber can not be made from the "E" value alone, however; its "F_b" or Extreme Fiber in Bending must also be considered.

Extreme Fiber in Bending: This means that lumber is assessed according to its fiber bending and resisting properties. "Extreme fiber stress in bending" occurs when loads are applied, producing tension in the fibers along the face farthest from the applied load and inducing compression in the fibers along the face nearest to the applied load. These "F_b" values have been measured in laboratory tests and are assigned numerical quantities. It is important to understand their role in grading lumber.

Refer to Table 3–1 again and note the different grades of lumber listed on the left, starting with Dense Sel Str KD. You see that the "F_b" value is rated from 2200 psi to 2500 psi. A board graded accordingly would have a Modulus of Elasticity of 1.9 million psi. A

TABLE 3–1:   ALLOWABLE STRESS BY GRADE OF LUMBER *(Courtesy of Southern Forest Products Association)*

GRADE	Extreme Fiber In Bending "F_b"*		Modulus of Elasticity "E" psi
	2-4" THICK, 5" & WIDER psi	2-4" THICK, 2-4" WIDE psi	
Dense Sel Str KD***	2200	2500	1,900,000
Dense Sel Str	2050	2350	1,800,000
Sel Str KD	1850	2150	1,800,000
Sel Str	1750	2000	1,700,000
No. 1 Dense KD	1850	2150	1,900,000
No. 1 Dense	1700	2000	1,800,000
No. 1 KD	1600	1850	1,800,000
No. 1	1450	1700	1,700,000
No. 2 Dense KD	1550	1800	1,700,000
No. 2 Dense	1400	1650	1,600,000
No. 2 KD	1300	1550	1,600,000
No. 2	1200	1400	1,600,000
No. 3 Dense KD	875	1000	1,500,000
No. 3 Dense	825	925	1,500,000
No. 3 KD	750	850	1,500,000
No. 3	700	775	1,400,000
Construction KD		1100	1,500,000
Construction		1000	1,400,000
Standard KD		625	1,500,000
Standard		575	1,400,000
Utility KD		275	1,500,000
Utility		275	1,400,000
Stud KD	800**	850	1,500,000
Stud	725**	775	1,400,000

*See section entitled "Design Values." The extreme fiber stress in bending, F_b, was increased by 15% to 1.15 F_b for repetitive member uses.
**Applies to 5" and 6" widths only.
***Terms and abbreviations. Sel. Str. means select structural; KD means KD15, dried to a moisture content of 15% or less; where KD is not shown the material is dried to a moisture content of 19% or less. Lumber dried to 19% or less will be stamped S-DRY or KD S-DRY.

"1.15 F_b" listing means that there is an allowable variation of 15% provided that 3 or more joists or members are uniformly spaced at not more than 24" oc and are joined by flooring or other load-distributing material.

Appendix A at the end of this book has a series of tables for use with several types of construction lumber. The calculations tabulated in them provide data according to grade, size, "F_b" and "E". To verify requirements for a particular job you should refer, when necessary, to these tables. (See the note in Appendix A for a full listing of all species of lumber.)

Let's work several problems to appreciate how this data relates to structural standards. What you should understand is the application of the terms just learned to the factors important for selecting stock that can support loads safely.

PROBLEM 1: Verify the design criteria of an office floor unit planned for construction from No. 2 Dense KD 2 × 8's with a span of 11'6" between girder and wall and a spacing of 24" oc.

Figure 3–2, an excerpt from the tables included in Appendix B, lists the allowable spans for floor joists designed for a 40 psf live load.

Selecting Joist Size: Read across the headings at the top of the table until you locate the grade of lumber specified. You will find No. 2 Dense KD listed at the right, in the fifth column (or fourth after the table's so-called "stub" at the far left). Looking down the column you can see that no 2 × 6 would be strong enough to do the job. Any on-center joist spacing for 2 × 8's would do, however, and the best, of course, being the 24" oc.

Choosing Grade: Now let's examine how grading is chosen. Again refer to Table 3–1 and see that No. 2 Dense KD lumber grade is assigned an "Fb" of 1550 psi and an "E" of 1.7 million psi.

Selecting the Species: Refer to Appendix A (page 241). On page 245 locate Southern Pine. But, note that the heading states "Surfaced dry" which indicates a 19% moisture content. Recall, however, that the problem specifies "KD" which means Kiln Dried. In other words, you need to look at other data on page 245 which lists species of lumber "surfaced at 15% moisture content, KD."

Determining Quality of Species: Next, for 2 × 6 and wider stock, trace the data for Dense No. 2 and compare quality and grade.

Table 3–1   "F_b" = 1550 psi; "E" = 1.7 million psi

Table W–1   "F_b" = 1750—2190; "E" = 1.7 million psi

CONCLUSION: A Southern Pine 2 × 8 of grade No. 2 Dense KD will provide adequate support at 40 psf live load.

Now that you understand the method, let's evaluate data for another problem, using the same figures and tables but under different conditions.

Size and Spacing in. o.c. / Grade²	Dense Sel Str KD and No.1 Dense KD	Dense Sel Str KD, No.1 Dense and No.1 KD	Sel Str, No.1 and No.2 Dense KD	No.2 Dense, No.2 KD and No.2	No.3 Dense KD[3]	No.3 Dense[3]	No.3 KD[3]	No.3[3]
2 x 6 12.0	11-4	11-2	10-11	10-9	10-1	9-9	9-4	9-0
13.7	10-10	10-8	10-6	10-3	9-5	9-2	8-9	8-5
16.0	10-4	10-2	9-11	9-9	8-9	8-6	8-1	7-10
19.2	9-8	9-6	9-4	9-2	8-0	7-9	7-4	7-1
24.0	9-0	8-10	8-8	8-6¹	7-1	6-11	6-7	6-4
2 x 8 12.0	15-0	14-8	14-5	14-2	13-3	12-11	12-4	11-11
13.7	14-4	14-1	13-10	13-6	12-5	12-1	11-6	11-1
16.0	13-7	13-4	13-1	12-10	11-6	11-2	10-8	10-3
19.2	12-10	12-7	12-4	12-1	10-6	10-2	9-9	9-5
24.0	11-11	11-8	11-5	11-3¹	9-5	9-1	8-8	8-5
2 x 10 12.0	19-1	18-9	18-5	18-0	16-11	16-5	15-8	15-2
13.7	18-3	17-11	17-7	17-3	15-10	15-5	14-8	14-2
16.0	17-4	17-0	16-9	16-5	14-8	14-3	13-7	13-1
19.2	16-4	16-0	15-9	15-5	13-5	13-0	12-5	12-0
24.0	15-2	14-11	14-7	14-4¹	12-0	11-8	11-1	10-9
2 x 12 12.0	23-3	22-10	22-5	21-11	20-7	20-0	19-1	18-5
13.7	22-3	21-10	21-5	21-0	19-3	18-9	17-10	17-3
16.0	21-1	20-9	20-4	19-11	17-10	17-4	16-6	16-0
19.2	19-10	19-6	19-2	18-9	16-3	15-10	15-1	14-7
24.0	18-5	18-1	17-9	17-5¹	14-7	14-2	13-6	13-0

1. The span for No. 2 grade 24 inches o.c. spacing is: 2x6, 8-4 2x8, 11-0 2x10, 14-0 2x12, 17-1.
2. Terms and abbreviations: Sel Str means select structural. KD means KD15, dried to a moisture content of 15% or less, where KD is not shown the material is dried to a moisture content of 19% or less. Lumber dried to 19% or less will be stamped S-DRY or KD S-DRY.
3. These grades may not be commonly available.

Figure 3–2 Data for Floor Joists Carrying 40 psf Live Load (excerpt from Table 2 in Appendix B)

PROBLEM 2: Select a species of lumber suitable for framing a floor unit to meet the following specifications:

a) Span is 14'9", joists on 24" oc.

b) "E" at 1.6 million psi or greater.

c) "F_b" at 1400 psi to 1650 psi.

Available Grades: From Table 3-1 we learn that several grades may be considered. From the minimum grade to best in terms of allowable stress they are as follows:

a) "F_b" = 1400—1650 "E" = 1.6 million psi = No. 2 Dense

b) "F_b" = 1550—1800 "E" = 1.7 million psi = No. 2 Dense KD

c) "F_b" = 1450—1700 "E" = 1.7 million psi = No. 1

Other grades are significantly better.

Choosing Grades: Keeping in mind that the span is 14'9" at 24" oc, look again at Figure 3-2 and find the grades in the table boxheads across the top. Listed below are some of the joists whose ratings correspond to one you would consider:

a) No. 2 Dense: a 2 × 10 would barely do; the span is 14'9" and the table lists 14'7" as the normal limitation.

b) No. 1: a 2 × 10 would be adequate; in addition, the "F_b" and "E" qualities would be better.

Selecting the Species: Return again to Appendix A (pages 242 and 245), recalling that the requirements are for Dense No. 2 KD grade with "F_b" = 1550 psi, "E" = 1.7 million psi, minimum 2 × 10.

a) Douglas Fir—Larch; excellent choice "F_b" = 1700, "E" = 1.7 million psi. (Note that surfaced dry @ 19% moisture still exceeds our needs.)

b) Douglas Fir South; *not* acceptable.

    1) fiber bending quality is inadequate; Modulus of Elasticity totally inadequate

2) grade is not listed.

c) Southern Pine; acceptable. It meets all requirements including KD.

CONCLUSION: The species that would provide the required support adequately with the specifications imposed by the plans is No. 2 Dense Southern Pine; however, if a variance could be obtained, Douglas Fir–Larch surface dried would be more than adequate and perform well.

These brief examples should provide you with an insight to a better understanding of lumber and its structural function.

Sheathings General Qualifications

Sheathing materials also have an important role in construction. Although builders still require sheathing lumber, such as 1 × 6's or 1 × 8's "s4s" and T&G, most specify plywood and fiberboards which have generally replaced the lumber. Several factors have brought about this change. Pricing is a major factor. The cost of ¾ " lumber and the cost of the required nailing as well as the cost of wall bracing material make lumber sheathing very expensive.

Plywood and fiberboard panels, although somewhat expensive initially, do reduce overall costs. For example, an average panel or sheet contains 16 to 32 square feet, can be installed quickly, and eliminates the need for wall bracing in many instances. What makes these products so desirable? Primarily their ability to restrain *racking* pressures and forces and the variety of design features.

Racking or forcing a frame out of plumb can be reduced sharply with plywood. As you will read shortly, the sheathing applied in the 24" system of framing provides the stability necessary to withstand severe wind pressure as well as settling forces. When plywood sheets are nailed along the edges at 4" oc and on intermediate studs or joists at 6" to 8" oc, a very strong panelized structure is formed.

Plywood Grading for construction is done within strict quality parameters established by the construction industry and published as "Product Standard 1–83". Standardized tests have been developed in order to identify plywoods by groups, grades, thicknesses, and uses as an aid to both designers and builders. These ratings are stamped on all plywood panels. The collective data that determines the grading can be seen in Table 3–2. Let's identify several significant points in the identification index that tells us which plywoods are useful in framing and sheathing.

Table 3-2: Span ratings for sheathing and single-floor panels (Source: PS 1-83, P/O Table 6, p 20.)

Sheathing panels (C-D, C-C)				Single-floor panels (Underlayment, C-C Plugged)	
Span	Thickness (in inches)	Span	Thickness (in inches)	Span	Thickness (in inches)
12/0	5/16	40/20	5/8,	16 o.c.	1/2 ,
16/0	5/16, 11/32		21/32, 3/4, 25/32		19/32, 5/8
20/0	5/16, 11/32, 3/8	48/24	3/4, 25/32, 7/8, 29/32	20 o.c.	19/32, 5/8, 23/32, 3/4
24/0	3/8, 13/32, 1/2			24 o.c.	23/32, 3/4, 7/8
32/16	1/2, 17/32, 5/8			48 o.c.	1⅛

Span: left number for rafter o.c. spacing, right for floor joist spacing.

1) *Groups of Plywoods:* There are five in which 73 species of trees are listed. Some of the more widely used woods in each of the four major groups are listed below:

Group 1:
Birch, Sweet
Birch, Yellow
Douglas Fir 1
Larch, Western
Maple, Sugar
Oak, Tanbark (Tanoak)
Pine, Loblolly (Swamp)
Pine, Longleaf (Georgia)
Pine, Shortleaf
Pine, Slash (Caribbean)
Pine, Southern*

Group 2:
Cedar, Port Orford
Douglas Fir 2
Fir, California Red
Fir, Grand
Fir, Noble
Fir, Pacific Silver
Fir, White
Hemlock, Western
Lauan, Almon
Lauan, Bagtikan
Lauan, Red

---

*"Southern Pine" is sometimes used to designate all the pine trees listed here, including the Carribbean pine since it grows in Florida, as well as in Cuba and the Bahamas.

Group 2 (cont.):

Lauan, Tangile	Pine, Lodgepole
Lauan, White	Pine, Ponderosa
Maple, Black	Pine, Spruce
Menkulang	Redwood
Meranti, Red	Spruce, Engelmann
Pine, Pond	Spruce, White
Pine, Red	Hemlock, Eastern
Pine, Western Pitch	*Group 4:*
Pine, Virginia	Aspen, Bigtooth
Spruce, Sitka, Red, Black	Aspen, Quaking
Sweetgum	Cedar, Incense
Tamarack	Cedar, Western Red
Yellow Poplar	Pine, Eastern White
Group 3:	Pine, Sugar
Alder, Red	*Group 5:*
Birch, Paper	Basswood
Cedar, Alaska	Poplar, Balsam
Jackpine (Gray)	

2) *Grades* are assigned to plywoods according to appearance quality and structural quality. The better (more clear of defects, for instance), the higher the alpha grade or rating. Appearance grades are assigned as "N" and from "A" through "D": grades "N" and "A" are veneers suitable for staining and painting; "B" has up to 5 percent defects; "C" has discolorations and sound knots up to $1\frac{1}{2}''$ across the grain; "D" has knots up to $2\frac{1}{2}''$ across and has been patched as required. The structural grades are STRUCTURAL I C-D unsanded, STRUCTURAL I C-D Plugged Underlayment touch sanded, STRUCTURAL II C-D unsanded, and STRUCTURAL II C-D plugged Underlayment, touch sanded.

3) *On-Center Space Limitations versus Thickness* are also important in commercial construction. Table 3-2 lists most of the standard thicknesses of sheathing plywood panels. Typical classifications from Table 3-2 can help solve the problems in the two cases we examine next.

CASE 1: Joists on 24" oc, C-D exterior or structural I C-D, flooring sheathing. What is the minimum thickness of plywood that could be safely used?

USE OF DATA: Read the values in the columns under the first major heading for C-C Exterior and C-D Interior sheathing. That is, all numbers on the left of the slash mark (/) indicate rafter spacing; all numbers to the right indicate joist spacings; and a zero (O) means not acceptable for use as rafter or joist. A careful examination of data reveals the following possibilities:

$\frac{3}{4}$"   C-C Exterior and Structural I or II C-D.

$\frac{25}{32}$"   C-C Exterior and Structural I or II C-D.

$\frac{7}{8}$"   C-C Exterior and Structural I or II C-D.

CONCLUSION: Either $\frac{3}{4}$" Group 1, Douglas Fir or Pine, for instance, or $\frac{7}{8}$" Group 2 or 3 Fir, Pine or Spruce would be acceptable floor sheathing for Case 1.

CASE 2: Underlayment for a single-floor panel made from plywood can be one of seven panel thicknesses. What thickness is applicable to the requirements for decking a floor with joists at 24" oc?

USE OF DATA: From right hand column of Table 3-2

24 o.c. $\frac{23}{32}$"
              $\frac{3}{4}$"
              $\frac{7}{8}$"

CONCLUSION: Several underlayment thicknesses are suitable, so pricing would probably be the deciding factor.

Fiberboard has many of the same qualities that make plywood so useful for sheathing walls. The vertical installation of sheets nailed properly with 3" spacing around the edges and with 6" spacings on intermediate members results in a rigid form. This is especially true of the fiberboard manufactured for the purpose, such as insulating sheathing. This grade eliminates the need for corner bracing and is able to hold nails driven through siding. (See Figure 3-3).

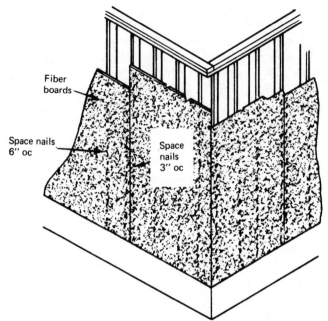

Figure 3–3 Method for Installing Fiberboard

Significant information on structural properties comes from this study of the materials used for framing. We now understand the reason for selecting a grade of frame lumber able to carry the predicted load and a sheathing necessary to a quality job. With this in mind, let's review general framing techniques and then study the 24-inch system. Then we can understand how soundproof wall systems can be developed with these building techniques.

Methods and Procedures

Figure 3–4 illustrates the customary framing practices used for almost every type of building. They include the building of girders, joist units, exterior and interior walls and partitions, ceiling joists, as well as sheathing floor tasks. As a rule, the procedures followed in residential construction apply in commerical carpentry, except that mass production methods are frequently used for cutting members and for pre-forming walls. Fastening, bracing, and stiffening methods are essentially the same. But closer spacing of members, fasteners, and stiffeners may be required because commercial building functions often involve heavier loads. A contractor would consider that a primary detail for attention during planning and estimating.

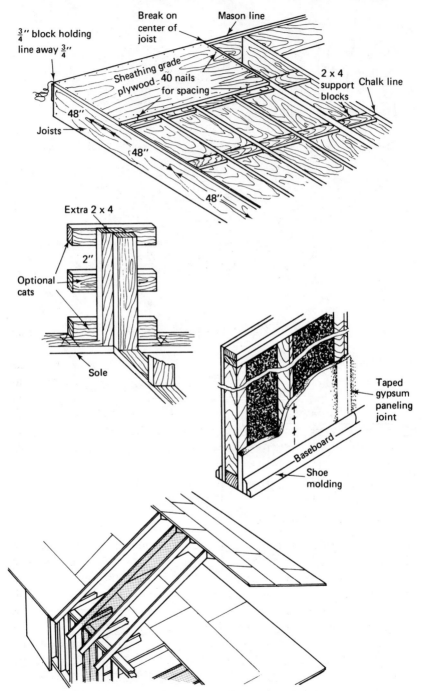

Figure 3–4 Customary Framing Styles

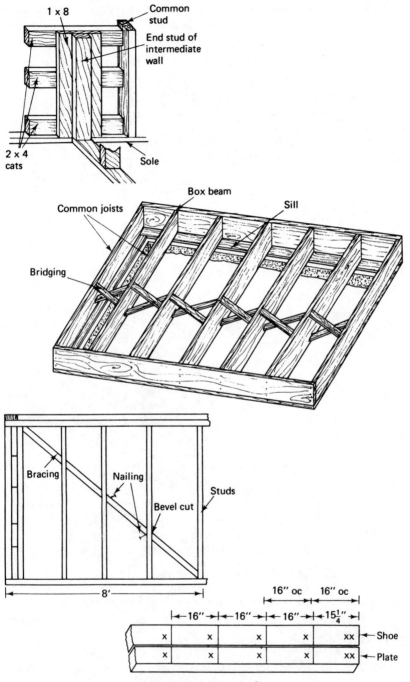

Figure 3-4 (continued)

If customary framing practices are properly understood, they can be adapted to various framing systems so as to save labor and materials, increase profits, and provide an actual gain in reliability. The "MOD-24 System" is one such system.

The MOD-24 System

The results of a study conducted by the National Association of Home Builders Research Foundation, Inc., showed that the 24-inch framing system could save 15.2% on labor for wall construction and 12.3% on labor for floor construction. In addition, there was a savings of approximately 6% in the cost of materials for framing. Since the MOD 24 system offers an opportunity for additional profit as well as flexibility and reliability, let's examine it briefly.

The system consolidates a series of important construction advances made over a period of years. Most elements in the system are already firmly established in building codes, regulatory standards, and construction practices. MOD 24 is based on the principle of alignment of standard-measure members, coupled with the use of plywood and fiberboard sheathing to provide racking stability. Joists, studs, and rafters are aligned in a series of modules or in-line frames. Figure 3–5 attempts to illustrate this by animation. Each frame resembles a band of steel. Stresses, forces and weights are distributed throughout the members. These are evenly spaced 24" oc, hence the name for the system. Soleplates, ceiling plates, and bridging provide stability and a method for fabrication; however, the sheathing and/or siding whether single- or double-wall, single- or double-flooring, and single-layer roof sheathing construction provide the stiffening and uniformly tie the frames into modules.

Floors: Floor framing should be constructed with the joists butted, if required, so that the ¾" plywood can be glued and nailed intact—without the necessity for cutting or modifying the panels. See Figure 3–6. Whether single or double underlayment is used, the edges and ends of each sheet must be nailed to solid members, joists or blocking.

Walls: They are assembled in a conventional manner. If the design of openings are carefully considered, however, the need for cripples and double studs is minimized. Figure 3–7 shows the results of such planning.

Layouts: They must be made from a common reference point in order to use the 24-inch system successfully in construction. Once that point is established, all joists, studs, and rafters are laid 24" oc.

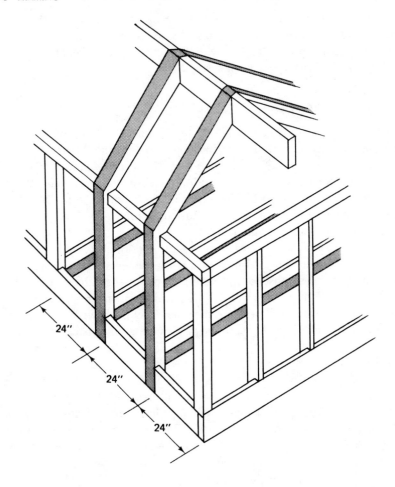

Figure 3–5 24-Inch System of Framing

A common reference point should also be established so that end walls and partitions can also be treated alike.

Conclusion: If the 24-inch system is adapted for use on your job, it follows that what you have learned about framing member grades and strengths and about plywood sheathing is directly relevant. These matters are integral to planning and will help you understand the principles involved. That in turn will ensure that reliability is built into any structure you work on.

In-line floor joists

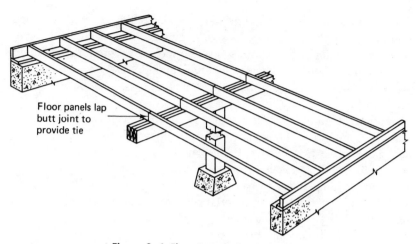

Floor panels lap
butt joint to
provide tie

Figure 3—6 Floor Preparation Technique

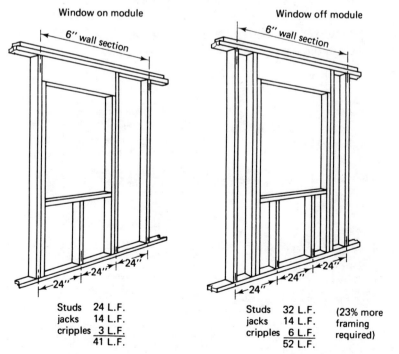

Window on module

6″ wall section

Window off module

6″ wall section

Studs	24 L.F.
jacks	14 L.F.
cripples	3 L.F.
	41 L.F.

Studs	32 L.F.	(23% more
jacks	14 L.F.	framing
cripples	6 L.F.	required)
	52 L.F.	

Figure 3—7 Layouts for Window on Module and for Window off Module

THE SOUNDPROOF WALL AND FLOOR

Now let's examine how sound proofing can be achieved through design and construction of framing. Design engineering, especially for townhouse and commercial buildings, usually includes some form of sound control. By carefully selecting materials and fabrication methods unwanted airborne, structural-borne, and impact noises can be effectively reduced.

Sound at audible levels (from 125 to 4000 cycles per second) is radiated through the structural members, plywood paneling, and gypsum panels. To be effective then, sound proofing must reduce such sounds to inaudible levels.

The building industry has developed a set of values based upon the speech carry-through of material substances, such as walls or floors. Accordingly the sound-deadening qualities of certain combinations of wall and floor materials are rated according to the way they can absorb radiated sound in predictable measures. For example, a barrier with an *STC* (sound transmission class) of 50 would be adequate for a partition between two rooms. See Figure 3–8-A. Note that the calculations charted in Figure 3–8-B show that materials having an *STC 50* value reduce loud speech to an inaudible level. Study the other levels in the data carefully.

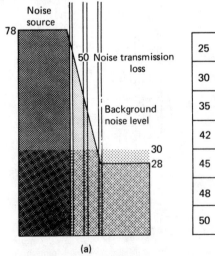

25	Normal speech can be understood quite easily
30	Loud speech can be understood fairly well
35	Loud speech audible but not intelligible
42	Loud speech audible as a murmur
45	Must strain to hear loud speech
48	Some loud speech barely audible
50	Loud speech not audible

(a) (b)

Note: A good sound barrier should reduce the noise originating in one room to below the background noise level in an adjacent room.

Note: This chart from the Acoustical and insulating Materials Association illustrates the degree of noise control achieved with barriers having different STC numbers.

Figure 3–8 Audible Levels Ratings *(Courtesy of Western Woods Products Association)*

Keeping this data in mind, let us examine several cases that use various wall and floor structures.* The objective is to understand which construction methods and materials combine best for sound deadening in walls or floors.

CASE 1: A single-stud wall constructed of 2 × 4 studs with double top-plate and a single bottom-plate or sole.

MATERIALS DATA:

SOUND CONTROL FACTORS	RATING
Gypsum panel applied with screws on both sides.	STC 34
Gypsum panel laminated and nailed over sound board on each side.	STC 45
Single gypsum board applied with screws one side, opposite side on resilient channels and 1½ "glass film insulation.	STC 50

CONCLUSION: Single-wall construction techniques offer very little sound-deadening. Only when sound board or resilient channels and insulation are used does noise absorption improve so that the wall qualifies as sound controlled. (Note: more on this in Chapter 6 on interiors.)

CASE 2: Double-stud wall with common plates and 2 × 3 or 2 × 4 studs—each offset by 2" to 8" to prevent any chance of contact.

MATERIALS DATA:

SOUND CONTROL FACTORS	RATING
Single gypsum board each side, applied with screws, no resilient channels, but 2" mineral wool installed.	STC 49
Single gypsum board laminated and nailed over sound deadening board each side, no resilient channel or mineral wool.	STC 49
Single gypsum board nailed on side. Single gypsum board on the resilient channels opposite, plus 1½ " glass fiber insulation.	STC 50

CONCLUSION: Double-wall construction that uses a common plate and soleplate qualifies as well sound-controlled with the addition of certain insulating materials.

---

*Data for the five cases presented here are excerpted from material on sound control published by the Western Woods Products Association.

CASE 3: Double-stud walls on separate plates consisting of 2 × 3 studs with plates 1″ apart and with studs offset 2″ to 8″.

MATERIALS DATA:

SOUND CONTROL FACTORS	RATING
Single gypsum board applied with screws and 2″ mineral wool.	STC 51
Single gypsum board laminated and nailed over sound board on each side; no wool or glass insulation.	STC 53
Same as above but with 3″ layer of wool installed in walls.	STC 60

CONCLUSION: Double separated walls offer the best sound-deadening method for wall construction; however, its cost may be nearly double that for other kinds of walls.

Observations

From the cases just examined, we can fairly conclude that each type of wall can be made an effective sound-control device with the inclusion of mineral wool, fiberglass and resilient channels. Double layers of gypsum board also aid in sound control; and values as high as STC 59 and 60 are possible when it is used with mineral wool.

CASE 4: Conventional wood-framed floor with 2 × 10 joists, ½″ subfloor, and gypsum board applied with screws on resilient channels for ceilings.

MATERIALS DATA:

SOUND CONTROL FACTORS	RATING
⅛″ vinyl asbestos tile on ⅜″ plywood underlayment, no glass fiber or wool insulation.	STC 37
.075 vinyl sheet on ⅜″ plywood underlayment, plus 3″ glass fiber.	STC 46
Carpet and pad directly over floor, plus 3″ glass fiber.	STC 47
$2^5\!/_{32}$″ oak strip over subfloor plus 3″ mineral wool.	STC 50
Carpet and pad added to floor over oak strip and 3″ mineral wool.	STC 50

CONCLUSION: Insulation such as 3″ glass fiber or mineral wool, is required to make this structure sound deadening. The use of wood-strip flooring over the subflooring adds very little sound control if the rooms are to be carpeted anyhow.

CASE 5: Conventional wood frame with ⅝" subfloor, ½" sound board, ½" underlayment, 3" glass fiber insulation, ⅝" gypsum on resilient channels.

MATERIALS DATA:

SOUND CONTROL FACTORS	RATING
⁵⁄₁₆" woodblock (parquet) as finished floor.	STC 54
Carpet and pad as finish floor.	STC 55
Vinyl finish flooring laminated to underlayment.	STC 58

CONCLUSION: Inclusion of sound-deadening board, mineral wool, and underlayment in the basic floor structure are the greatest sound-controlling factors. Addition of finish flooring, whether wood, rug or tile accounts for no more than a *4 to 8 STC* difference.

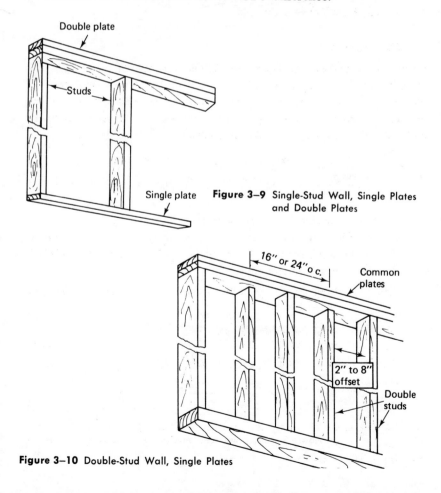

Figure 3–9 Single-Stud Wall, Single Plates and Double Plates

Figure 3–10 Double-Stud Wall, Single Plates

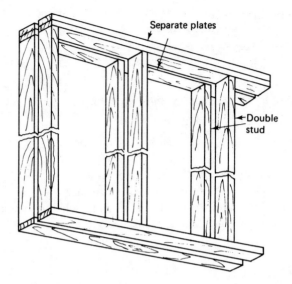

Separate plates

Double stud

Figure 3–11 Double-Stud Wall, Separate Plates

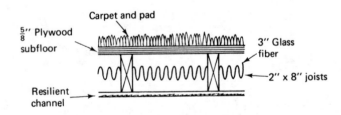

Carpet and pad

$\frac{5}{8}''$ Plywood subfloor

3'' Glass fiber

2'' x 8'' joists

Resilient channel

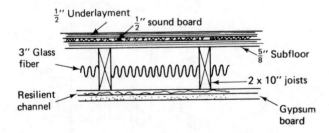

$\frac{1}{2}''$ Underlayment

$\frac{1}{2}''$ sound board

3'' Glass fiber

$\frac{5}{8}''$ Subfloor

2 x 10'' joists

Resilient channel

Gypsum board

Figure 3–12 Layouts for Conventional Floor

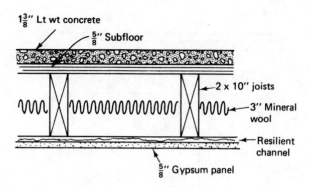

Figure 3–13 Layout for Modified Conventional Floor

Overall Conclusion

1. Sound-deadening requirements imposed by the contract should be studied very carefully. Their requirements will make a significant impact on both cost of material and cost of labor calculations.

2. The degree of sound proofing specified (*STC* quality) will, in part, dictate the wall and floor construction method to use.

3. Where options are available, a cost analysis must be made to specifically include the variables of material, time for installation, availability of materials and training requirements.

The wood-framed partition is accepted by all the building codes, and, as you just learned, it can be made soundproof if properly insulated and covered.

METAL PARTITIONS

Many times, though, the need arises for a more nearly fireproof and light-weight wall system. Then a metal framing unit often provides the answer. Metal stud partitions are light weight, non-load bearing assemblies. They are fabricated by U.S. gypsum and others from steel channel studs set in floor and ceiling runner tracks and may be faced with gypsum panels. The studs and metal runners are available in widths from $1\frac{5}{8}''$ to $6''$. Figure 3–14 shows a partition including a door opening. Note the way in which the metal is cut around the head of the door.

Metal partitions are usually soundproofed in the same way as

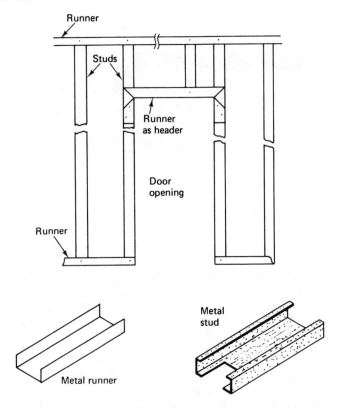

Figure 3–14 Details of Metal Non-Load Bearing Partition

wood partitions. The use of 3″ mineral wool and sound-deadening wallboard plus gypsum panels establishes a quality *STC* rate of 50.

Metal "bucks" consist of door jamb units constructed of pre-formed metal and joined channels. They were originally designed for use in concrete walls; however, several varieties are now also used with conventional wood frame or metal stud partitions. See Figure 3–15.

The type used in concrete walls is usually set in place during wall construction. The blocks are then cemented around it; or it may also be set into a wall form before concrete is poured. Figure 3–16 shows a typical usage. Notice that 2 × 4 or similar stock is used to hold the jamb in position while the masons are at work.

The other types of bucks for use with wood frames and metal partitions come pre-assembled. They are assembled while being installed. They are screwed to wood or metal studs. Then standard procedures are followed to make them plumb and level. Separation of

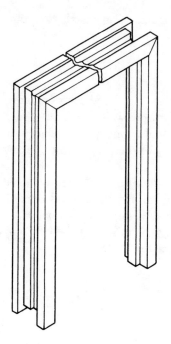

Figure 3–15 Frame Bracing for Concrete Walls

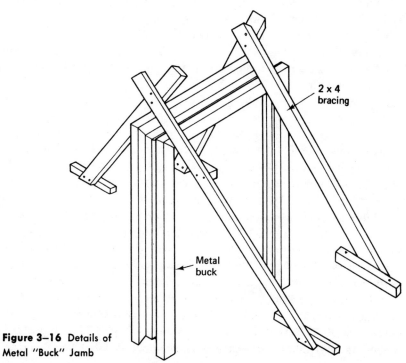

2 x 4
bracing

Metal
buck

Figure 3–16 Details of
Metal "Buck" Jamb

jamb sides and head are done as for regular jamb installations and according to manufacturer's suggestions.

Storefront Framing

Modern storefronts frequently consist mainly of large glass windows inserted in aluminum channels fastened in low walls. As a rule, such installation work is subcontracted. The usual framing requirement for the carpenter, therefore, is the preparation of the proper size opening detailed in the plans.

Adequate fastenings are usually available if the wall is constructed of wood. If the wall is concrete, however, the blocks of wood remaining after the forms have been removed may be all there is on which to fasten the windows. But frequently, anchor bolts are installed directly in the concrete or block by the subcontractor, so there may be little work for the carpenter.

TRAINING IN FRAMING

Training in framework usually requires home study by those new to commercial carpentry since it is important to learn thoroughly the theory that supports daily practice. Be sure to have trainees concentrate their studies on the principles underlying good framing methods. The chart you make for a training program must contain all the elements shown in Figure 3–17.

Record the Training

Prior to the actual starting date of the framing phase, you should review the Training Chart you have already made. The purpose for the review is to define any training needed on the theory of framing and identify specifically those techniques relevant to the current job. Since you have studied the appropriate organizational functions and have made your plans, the training requirements should be well known. They will include many of the task activities detailed on pages 90–95 for the improvement of framing techniques. Some of the most significant are: floor and ceiling layout, wall layout, cutting and assembly of framing members, and ways to reinforce, level, and plumb all work.

For the journeyman who needs to broaden his knowledge and skill in certain areas, such as the 24-inch system or the metal stud installations, training through task study must be obtained before practical

Name of Employee	Skill	Sills and girders	Floor joists and bracing	Stud layout and erection	24 inch system general	Bracing of walls	Corner preparation	Ceiling joist	Ceiling corner prep.	Floor or ceiling openings	Window and door module framing	Remarks

Basic (hatched) Skilled ⊠ Apprentice Ⓐ

Some (filled) Specialist ☐ None ☐

Figure 3–17 Training Chart Record of Framing Crew

application can be successful. Beginners and apprentices should be particularly encouraged to improve their knowledge and skill before the actual need arises. The exercises in this text may well be sufficient.

Framing Task Activities

Several task activities coded FR are presented and developed in much the same way as those in Chapter 2.

FR 1 TASK ACTIVITY: SILL, JOIST AND SUBFLOOR INSTALLATION

RESOURCES

Estimated Manhours for 2 men: 8 hrs per 400 sq ft floor area

Materials:

floor plans
joists
bridging—solid, cross, metal
sills
nails—16d common & 8d common
underlayment (APA approved grade & type)

Tools:

#8 crosscut saw	6-ft folding ruler
framing square	16-oz hammer
chalk line	combination square
7¼" power saw or	50-ft tape
radial arm saw *(optional)*	

PROCEDURES

1. Position sills for marking and for drilling, drill and insert sill over the anchor bolts; secure. Toenail joints (end butts) in sills. Prepare and install the girder if required; tie to sills.

2. Prepare box beams (end headers) and install as required by plans. Lay out joist spacing by starting from one end at either 16" or 24" oc; pre-cut joists for length. *Butt* ends of joists at splices; for plywood underlayment subflooring *DO NOT OVER-LAP*. Install joists with *crown side up* and nail.

3. Cut and install blocking for the edge-nailing of underlayment. Cut and install the bridging.

4. Frame out any required openings; use double headers on each end of opening and double the joist on each side.

5. Install underlayment, using edge and end as a squaring guide. Allow the proper separation between sheets; break joint on succeeding rows. Nail plywood according to schedule.

FR 2 TASK ACTIVITY: PLATE INSTALLATION AND STUD LAYOUT

RESOURCES

Estimated Manhours for 1 man: 1 hr per 20 lin ft

Materials:

floor plan
2 × 4 soleplates and wall plates
#12d common nails

Tools:

#8 crosscut saw	16-oz hammer
50-ft steel tape	6-ft folding ruler
framing square	adjustable combination square
brace and bit	

stud driver for concrete floors (*optional*)
electric drill and spade bit (*optional*)

PROCEDURES

1. Select straight 2 × 4's for soleplates and wall plates; using floor plans, locate and set a soleplate where each exterior wall and partition is called out. Cut, fit, and secure soleplate to floor.

2. Position wall plate next to soleplate, starting at one corner.

3. Mark off stud placement with first stud separation at either 15¼″ or 23¼″ and the additional separations marked each 16″ or 24″ oc. Use framing square to make lines across both plates at each mark. Place an (✕) denoting stud placement next to each line.

4. Obtain window and door placements from plan; mark plates accordingly.
5. Cut off wall plate to a length that will center on a common stud.
6. Continue process (Steps 2–5) until all exterior walls are completed; then repeat process for all partitions.
   NOTE: Where MOD 24 or similar system is used, start layout of studs on exterior opposing walls from the same end of the building.

FR 3 TASK ACTIVITY: WALL ERECTION, SHEATHING AND INNER/OUTER CORNER PREPARATION

RESOURCES

Estimated Manhours for 2 men: wall erection 8 hrs per 200 lin. ft
sheathing 8 hrs per 1000 sq ft
corner preparation 15 mins each

Materials:

floor plans
double-plate 2 × 4's
pre-cut headers, cripple
  and, jack studs
nails

pre-cut studs
sheathing, subflooring
corner pieces 2 × 4's
1 × 8's blocking pieces

Tools:

16-oz hammer
level
chalk line
power saw or
  vibrating saw (optional)

#8 crosscut saw
6-ft step ladder
keyhole saw

PROCEDURES

1. Prepare all necessary outside and inside corners.
2. Position all common studs and corners to the marks on previously installed soleplate. Position previously marked wall plate across the ends of studs; nail to studs. Frame out all door and

window openings. In double-wall construction (exterior walls) sheathing may be cut, fitted, and nailed to wall while it is still laying on the deck (See Chapter 5). Raise the wall upright to the vertical; secure with braces. Nail studs to soleplate.

3. Repeat process (Step 2) on other outside walls; then raise the inside (partition) walls. Plumb all corners, using the level and temporary braces.

4. Cut and install double-wall plate over first plate. Break joints to ensure adequate stiffening.

5. Complete inner corner construction; making sure each corner has adequate, stiff backing for interior paneling or gypsum wall board.

FR 4 TASK ACTIVITY: CEILING JOIST INSTALLATION AND CEILING CORNER PREPARATION

RESOURCES

Estimated Manhours for 2 men: 4 hrs per 400 sq ft ceiling area

Materials:

floor plans	pre-cut ceiling joists
bridging	2 X 4 blocking
1 X 8 or 2 X 8 corner boards	nails

Tools:

50-ft tape	6-ft folding ruler
combination square	framing square
#8 crosscut saw	16-oz hammer
ladder	

PROCEDURES

1. Pre-cut joists to required lengths (take measurement from floor plan); pre-cut bridging if it is to be used.

2. Lay out joist spacing to maintain 24″ system. Start marking from the same end of the building as for floor joists and studs.

3. Raise the joists to their approximate position on the walls with all crowns laying in the same direction. This standardizes the installation.

4. With a man working on either end of a joist, position joists and toenail to plates. After nailing all joists, align through the center of the span by using a 1 × 4 previously marked for on-center spacing.

5. Install any headers required for openings; follow by installing double joists either side of openings and short joists at intermediate on-center spacings.

6. Install bridging or other reinforcement as required.

7. Center and nail either a 1 × 8 or 2 × 8 over each partition that is parallel to ceiling-joist run; reinforce and stabilize piece by installing 2 × 4 blocking crosswise to the run and nailing into adjacent joists and down into either the 1 × 8 or 2 × 8.

FR 5 TASK ACTIVITY: SETTING METAL DOOR JAMB

RESOURCES

Estimated Manhours for 1 man: 25 mins per door

Materials:

floor plans	metal jamb
2 × 4 bracing stock	2 × 4 blocking
case-hardened nails (8d)	1 × 4 spreader stock
#8d common nails	

Tools:

16-oz hammer	#8 crosscut saw
level	6-ft folding ruler

PROCEDURES

1. Prepare one or two 2 × 4 blocks for jamb head and two to four 2 × 4 braces 8 ft long, or longer if the door is over standard height.

2. Position the jamb according to plan—usually on top of the concrete floor. Next, secure the base of each side jamb by fastening the floor knee to the floor while maintaining proper jamb separation; tacknail braces into block in head of jamb. Then, after

using the level to make sure that the jamb is plumb, secure the braces to floor.

3. The side jambs should be held in position by installing 1 × 4 spreader pieces near the base and half-way up.

FR 6 TASK ACTIVITY: CONSTRUCTING METAL PARTITIONS

RESOURCES

Estimated Manhours for 1 man: 2 hrs per 8 lin ft

Materials:

floor and detail plans
metal studs
metal runners
fasteners (screws, nails, and cement nails)

Tools:

stud and metal shears   6-ft folding ruler
metal lock fastener   14-oz hammer
snips
portable electric drill with screwdriver attachment

PROCEDURES

1. Measure, mark, and cut a pair of runners for a wall section; then measure, mark, and cut sufficient studs for one each at 24″ oc.
2. Lay out partition stud, spacing 24″ oc on both runners; then use metal lock fastener to connect studs to runners.
3. Form door openings by using same process (Step 2). Then cut a header to fit between side studs of jamb opening so sides of header strip extend over side studs. Using lock-fastener tool, crimp head in place.
4. Position the partition as required by plans and fasten to floor. Plumb studs, nail to ceiling joists or blocking between ceiling joists.

REVIEW QUESTIONS

1. What are some of the sources that may be used when calculating man-days for various subphases under production?
2. Define: Load, Deflection, Allowable Stress.
3. What does "40 lb per square foot live-load" mean?
4. Which deflection formula is applicable to joist selection for most commercial construction?
5. What does the term "modulus of elasticity" mean?
6. "Extreme fiber in bending" is a term used in _____.
7. Relate the choice of lumber species to the modulus of elasticity and fiber bending needed for a 2 X 8 floor joist with a 12-ft span.
8. What are the different grades of lumber?
9. What does "KD" stand for in reference to lumber?
10. Name the woods in Group I that are commonly used for lumber.
11. What thickness and grade of plywood would be needed to apply a subfloor across joists spaced 24" oc?
12. Explain the "24-inch framing system" briefly.
13. Why should the ends of floor joists be butted rather than over-lapped if plywood subflooring is to be installed?
14. "STC" means _____.
15. Define the two methods of framing a wall that aid in sound control.
16. Beside the steel stud what else is needed to construct a metal partition?
17. What is a metal "buck"?

4

Roofing Tasks and Techniques

Base leg: in a right-angled triangle the *adjacent* side or the side opposite the right angle; in a roof layout the horizontal rafter run

Bird's mouth: the right-angled notch cut in the underside of a rafter; designed so as to allow the rafter to seat properly on the wall plate

Chord: the principal member of a truss extending from end to end, usually one of a pair and connected by a web: the incline (rafter) and horizontal (joist) members

Connector plates: usually metal plates, used for anchoring ring truss assemblies to walls or columns

Crown: that edge of a rafter where the bow or curve in the joist can be seen as uppermost in the middle of its length

Flat roof: the simplest type of roof, has very little pitch (up to 3 in 12); used extensively in commercial construction

Hip: a type of roof having members that extend from a building corner so as to slope at an oblique angle to the ridge; also the rafters so cut to make this roof

Jack rafter: a rafter having less than the full length of the roof slope, as one meeting a hip roof; one end has a common rafter cut and the other has a compound miter cut

Monoplaner: a type of roof truss built with a single thickness of stock lumber

Overhang: the soffit or the underside of extensions beyond the exterior wall surface, usually consisting of a projecting portion of the rafters

Pythagorean theorem: used as the basis for calculating rafter length; a geometric principle which states that

$$AB = \sqrt{AC^2 + BC^2}$$

Ring truss: a roof truss assembly that incorporates rings of steel to secure joining members

"Stick-built" roof: the external upper covering of a structure built on-site from individual rafters, ridges, collar ties, and braces

Web: one of the internal, integral members of a truss assembly that connects the stronger parallel members or chords

As explained in Chapter 3, we will examine roofing separately. Whole books have been written about roofs, so this chapter cannot and is not designed to tell all about roof construction. It will, however, provide data on gable roofs, on hip and valley construction techniques, and on truss roofs. Also included in this study will be the application of sheathing and shingles and methods for handling designed overhangs finished with shingles and other exterior materials.

Thorough knowledge of this work phase and its related tasks combine to create a true understanding of the roof's importance and function. Contemporary architects place much emphasis upon the architectural beauty achieved through roof design. Frequently too, the roof is shaped in such a way that the external integration of roof and side wall are all but indistinguishable. The building of such structures demands whole-hearted attention from all members of the work force and a clear, in-depth understanding of proper roofing techniques.

On the opposite side of the coin, so to speak, there is the old time flat roof which is still in use today. This, of course, presents no obstacle to the builder. Nevertheless, a summary of its major features and its strong points will give to inexperienced or unskilled carpenters the basics from which they can work toward a more complete understanding of the building techniques involved.

Be reminded that working on roof construction means working aloft. Two safety measures must be stressed and provided for at all times. First, carpenters must practice safety when at work on scaffolding and roof jacks. They must stay aware of their fellow workmen on the ground and must take every precaution to avoid dropping excess materials to the ground unless spotters and supervisors are on hand. Secondly, crew members on the ground below should try to avoid working in the immediate vicinity when roofing construction activities are in progress. They should also wear hard hats and be alert to overhead work if they must enter the area.

As with other phases of the total work, the organizational activities in roofing focus attention on the responsibilities of contractor and staff. So in this opening section of the chapter we discuss the impact of the roofing phase of construction.

ORGANIZATIONAL FUNCTIONS IN ROOFING CONSTRUCTION

A very expensive part of any construction job is the building of the roof, so very careful planning by contractor and staff is essential. The five subphases listed under scheduling in Figure 4–1 only partly identify the complex of decisions necessary to implement a total plan. Good roof construction requires maximum concentration and effort of all personnel to devise and carry out a complete plan for its erection and completion.

Type of roof and installation techniques are among the most important factors in planning. Thus, if a truss roof is to be installed and

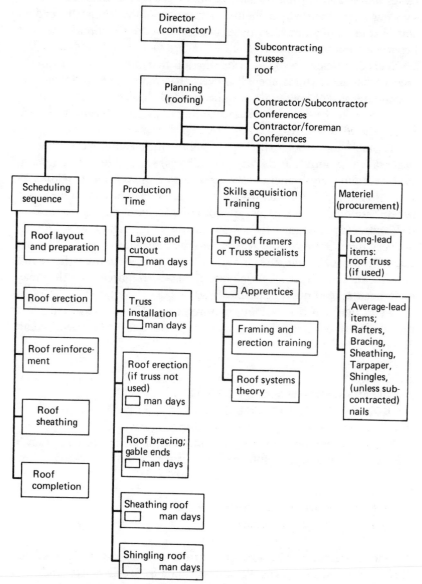

Figure 4–1 Organization Chart of Roofing Activities

the making of trusses is to be subcontracted, they must be considered long-lead items. Similarly, the actual installation of the trusses means that trained workmen are needed; therefore, training may be required and will have to be included in the plan.

The Figure 4–1 chart provides a logical break-down of the roofing phase of construction. Most of the activities for the job can logically be grouped under production. Man-days can be worked out by reviewing the estimated manhours in the task activities at the end of this chapter. They may also be calculated by taking approximate past experience into account.

This chart can also be used for any job that calls for a "stick-built" roof or flat roof. Remember, however, that more time, training, and a different list of materials will be needed. These decisions all influence the final make-up of the time-line plan that you must also set up for the roofing phase.

A final note about the chart; notice that conferences may be needed. To be sure that the subcontractor(s) fully understand responsibilities and requirements, the contractor should meet with him (them). Thus the contractor can be assured that he can rely on his subcontractor(s). After plans are formulated with the subcontractor, the contractor must pass the data on to his staff, including scheduled delivery dates, type of roof truss, and erection requirements. The contractor must also confer with his foreman, inform him of the plans, and tell him about any special techniques or requirements that may have been agreed on at other meetings.

With the plans for training requirements, materials, methods, and the like fairly well established, let's begin to study the details and descriptions of roofing techniques.

DETAILS AND DESCRIPTION OF ROOF CONSTRUCTION TECHNIQUES

Materials and methods for building roofs must be carefully chosen if the roof is to function as intended. Guesswork is not a method a contractor can afford to use. Proper judgments must be based on scientific fact and/or engineering data.

Selecting the Rafter

The same engineering data that went into selecting the joists (Chapter 3) is used to select rafters. Tables in Appendix B summarize industry standards for rafters and their loads. Modulus of Elasticity and Fiber in Bending factors are again primary for quality control of the lumber grade and species necessary for adequate support.

Recall that three ratings for stiffness are $/360$, $/240$, $/180$. It was stated in Chapter 3 that an $/180$ rating would be acceptable for rafters under some conditions. $/360$ and $/240$ are both acceptable since the load stressess placed on floor joists are usually considerably more than on the roof. By studying the tables in Appendix B you can see that certain members (block type) are limited to $/180$ and others are limited to their grading.

Let's work out two problems in order to understand the importance of selecting the proper timbers for rafters.

PROBLEM 1: Select a rafter to carry a 40 psf load whose pitch is 6 in 12 and will be used on a building with a 36-ft span. Spacing will be 24″ oc and rafters must have a deflection limitation of $/240$.

SOLUTION:

1. Rafter run = 18 feet or ½ building span.
2. At a rise of 6 in 12 the length of rafter needed (not including overhang) = 20'0″.
3. Only 2 × 12 grade Dense Sel Str can be used. See Table 9, Appendix B.
4. Dense Sel Str = "E" of 1.9 million psi, "F_b" of 2100 to 2400 psi.
5. Species meeting both grade and load standards is *Douglas Fir-Larch*. See Appendix A.

PROBLEM 2: Select a rafter by species and grade that will meet the following requirements. The building will have a span of 26 feet and an 8 in 12 rise. Predicted dead load will be 15 psf dead load. Allowable unit stress in bending (F_b) will not exceed 1200 psi under normal conditions nor 1500 psi with a seven day loading. Rafter spacing equals 24″ oc.

SOLUTION:

1. Rafter.run = 13 ft or ½ building span.
2. At a rise of 8 in 12 the rafter length (not including overhang) is 16.5 ft.
3. Several possibile grade choices (See Table 7, Appendix B):
   a) 2 × 8 Grade (Sel Str)
   b) 2 × 10 Grade (No. 2)
   c) 2 × 12 Grade (No. 3)

4. Species data for the three grades (See Appendix A):
   a) For 2×8 rafters Grade Sel Str, the acceptable species:
      * Douglas Fir—Larch (F_b, normal 2050 psi;
        7 day load 2560 psi)
      * Douglas Fir South (F_b, normal 1950 psi;
        7 day load 2440 psi)
      * Southern Pine (F_b, normal 2050 psi;
        7 day load 2560 psi)
      * Eastern Hemlock (F_b, normal 1750 psi;
        7 day load 2190 psi)
      * Eastern Spruce (F_b, normal 1500 psi;
        7 day load 1880 psi)
      * Engelmann Spruce (F_b, normal 1350 psi;
        7 day load 1690 psi)
      * Hemlock Fir (F_b, normal 1650 psi,
        7 day load 2060 psi) and several others.
   b) For 2×10 rafters No. 2, all the previous species are acceptable *except*:
      N/A Eastern Spruce (F_b, normal 1000 psi;
        7 day load 1250 psi)
      N/A Engelmann Spruce (F_b, normal 950 psi;
        7 day load 1190 psi)
   c) For 2×12 rafters Grade No. 3, *none* of the previous species are acceptable. Typical reason:
      N/A Douglas Fir-Larch No. 3 has F_b normal 850 psi and
        7 day loading 1060 psi.
5. Stock Selection: Acceptable grade and species include: the 2×8, Grade *select structural* Hemlock-Fir or better; or 2×10, Grade No. 2 Hemlock-Fir, Douglas Fir, or Southern Pine.

You can see from the methods used to solve these two problems that your decision can be based on sound engineering data. You need not make judgments without facts. Furthermore, most structural specifications include tolerance factors that allow you to select from a wide range of materials. In many cases this range gives a latitude that can mean several hundred to thousands of dollars of savings without a reduction in the standards required for a successful job.

Rafter Layout

The basic rafter layout is made by measuring first from the outer edge of the wall stud or sheathing to one-half the building span. Then, from this distance, one-half of the ridge board thickness is subtracted. Next, the required overhang is added. Two methods for measuring and cutting are frequently used as on-site rafter layouts: one with the framing square, the other with a tape measure and by calculating.

Of these two, the framing square is less accurate but more frequently used. As you will see when we analyze it, the inaccuracy develops as the carpenter steps-off his square along the rafter he is marking.

The measurement is critical to the subsequent fit of the layout. It is important, then, that the layout start and end correctly so that the exact rise per foot run is maintained.

Common rafter elements are illustrated in the diagram in Figure 4–2. It shows the basic rafter layout requirements that assure the proper rise per foot and accurate fit at ridge and plate. Points A and B are parallel; each is a plumb mark which, as you know, can be made easily with the framing square.

Observe particularly that the total length of the rafter minus overhang is from the vertical line or point A to the vertical line or point B. So the layout can be started either on the crown or underside of the rafter or through the center.

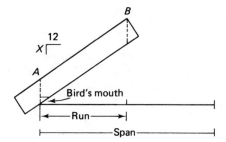

Figure 4–2 Common Rafter Layout

The *stepping-off method* is a step and repeat technique that is frequently used to step-off the rafter length in successive units. Briefly, the process involves a move or step of the framing square for each full foot of rafter run and a partial step for a fraction of a foot of the run. The connecting points are illustrated in Figure 4–3.

Once the ridge line is made, you move the square to the point where its 12-inch mark lays along the rafter edge. Mark it, then reposition the square at that point where the rise in inches (X) and the pencil mark are in line. Once again align the 12-inch marking on the square with the rafter edge and mark it. Repeat the process until the total run is measured off.

The distance Y is the length of rafter made with each step of the square.

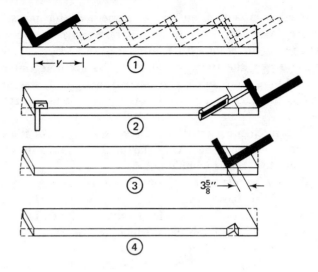

"y" times $\frac{run}{12}$ = rafter length.

Figure 4–3 Method for Stepping-off Rafter with Framing Square

Calculating the rafter length method uses the scale markings on the framing square. Those markings and Table 4–1 display the incremental arithmetic for the variety of rises per foot run necessary to calculate the rafter length. We shall use Figure 4–2 and the table again to illustrate several applications of this technique. During initial estimating of materials requirements this table or one like it is more convenient to use than a square.

PROBLEM 1: Building span equals 39'0"; plans call for 6 in 12 rise. Calculate the rafter length and lay out the rafter.

SOLUTION:

a) Dividing the span (39'0") by 2 gives the rafter run as 19.5 ft
b) Tracing across and down on Table 4–1 to find that for a 6 in 12 rise the length/ft run is *1 ft 1$\frac{42}{100}$ in.*
c) Multiplication of 19.5 ft × *1 ft 1$\frac{42}{100}$ in.* equals a total of *21 ft 9$\frac{68}{100}$ in.*
d) Layout procedure:
    1) Mark 6 in 12 ridge line
    2) Measure *21 ft 9$\frac{68}{100}$ in.* from either end of line along rafter edge and mark the point
    3) Mark a parallel 6 in 12 line at lower end of rafter

TABLE 4–1:   COMMON RAFTER LENGTHS PER FOOT OF RUN/RISE IN INCHES

Rise in inches	Incremental per inch run		Incremental per ft run			Angle		Rafter pitch
	in	100th	ft	in	100th	deg	min	
2	1 -	01333	1 -	0 -	16	9 -	30	1/12
3	1 -	03083	1 -	0 -	36	14 -	05	1/8
4	1 -	05417	1 -	0 -	64	18 -	25	1/6
5	1 -	08333	1 -	1 -	00	22 -	40	5/24
6	1 -	11833	1 -	1 -	42	26 -	35	1/4
7	1 -	1575	1 -	1 -	89	30 -	15	7/24
8	1 -	20167	1 -	2 -	42	33 -	40	1/3
9	1 -	25	1 -	3 -	00	36 -	55	3/8
10	1 -	30167	1 -	3 -	62	39 -	50	5/12
11	1 -	35667	1 -	4 -	28	42 -	30	11/24
12	1 -	41417	1 -	4 -	97	45 -	00	1/2
13	1 -	47417	1 -	5 -	69	47 -	15	13/24
14	1 -	53667	1 -	6 -	44	49 -	25	7/12
15	1 -	60083	1 -	7 -	21	51 -	20	5/8
16	1 -	66667	1 -	8 -	00	53 -	10	2/3
17	1 -	73417	1 -	8 -	81	54 -	45	17/24
18	1 -	8025	1 -	9 -	63	56 -	20	3/4
19	1 -	9375	1 -	11 -	25	57 -	45	19/24
20	1 -	96917	1 -	11 -	63	59 -	00	5/6
21	2 -	01	2 -	0 -	12	60 -	25	7/8
22	2 -	04167	2 -	0 -	50	61 -	5	11/12
23	2 -	08333	2 -	1 -	00	62 -	5	23/24
24	2 -	1115	2 -	1 -	38	63 -	5	1

Example 1: Rafter length for a 10 ft run with a 6 inch 12 rise:

a. Incremental per ft run for the rafter run is 1 ft 1.42 in. × 10 ft = 10 ft + 10 in. + $^{42}\!/_{100}$ths =

b. 10 ft + ($\frac{420}{100}$ = 4.20 in.) converting 100th to inches.

$$
\begin{array}{r}
10 \text{ in.} \\
4 \text{ in.} + {}^{2}\!/_{100} \\
\underline{10 \text{ ft} + 14 \text{ in.} + {}^{2}\!/_{100}} \\
\underline{+ \quad 1 \text{ ft} - 12} \\
11 \text{ ft} \qquad 2 \text{ in.} \qquad {}^{2}\!/_{100}
\end{array}
$$
subtracting 12 in. and adding 1 ft

Example 2: Rafter length for a 13 ft 4 in. run with a 4 in. 12 rise:

a. Incremental/ft run for the rafter run

1 ft $\frac{1.00}{100}$ in. × 13 ft = 13 ft + 13 in.

b. Incremental/inch run
1 in. .08333/100th × 4 in. = 4 in. + .3333

c. Add like quantities

$$
\begin{array}{r}
13 \text{ ft} \qquad 13 \text{ in} \\
4 \text{ in} \ .3333 \text{ in} \\
\underline{13 \text{ ft} \qquad 17 \text{ in} \ .3333} \\
\underline{+ \quad 1 \text{ ft} - 12 \text{ in}} \\
14 \text{ ft} \qquad 5 \text{ in} \ .3333 \quad = 14 \text{ ft} \ 5.33/100 \text{ in.}
\end{array}
$$
(converting 12 in. to 1 ft)

PROBLEM 2: The run of the building is 16'3" and the rise is 8 in 12. Calculate the rafter length with 12" overhang; lay out rafters for these requirements.

SOLUTION:

a) Mark an 8 in 12 line (ridge) along one end of the rafter so that the upper point of the line is on the crown of the rafter.

b) Using Table 4–1, read down to 8-inch rise data and then across to learn the length of the rafter per foot run.

c) Multiply 16 ft (run) by *1 ft 2*$\frac{42}{100}$ *in* and *3* in (run) by *1*$\frac{20}{100}$ in; add the results to obtain *19 ft 6*$\frac{34}{100}$ *in.*

d) Add overhang allowance of *1 ft 2*$\frac{42}{100}$ *in.*, making total length of rafter equal to *20 ft 8*$\frac{76}{100}$ in long.

Laying out Bird's Mouth: Almost every rafter requires a bird's mouth cut for proper seating on the plate. That in turn affords a secure method for nailing. Consisting of a heel cut which is vertical and a flat cut which is horizontal to the earth or plate, the bird's mouth notch also aids in rafter alignment during installation. Unless otherwise specified the flat part of the cut should be equal to or slightly longer than 3½" or the width of a 2×4 plate.

Overhangs: Whether 6 inches, 12 inches or wider, most rafter overhangs usually require a vertical cut and a horizontal cut. The vertical cut is the same as the ridge; the horizontal cut is made on the underside of the rafter and is determined by both the fascia width and the point where soffit and lookouts must attach to the wall. This point is usually several inches above window height.

Hip and Valley Principles of Layout

Once again you should remember that the framing square provides guidance for hip and valley construction. So that you understand the fundamental mathematical concepts that require the use of the 17-inch mark instead of the 12-inch mark, for instance, we will use Figure 4–4 and the Pythagorean Theorem. In Figure 4–4 you see a square whose sides are scaled to represent 24 inches. The center point (C) is one-half the distance of one side or 12 inches. This is the basis for the right triangle of which the common rafter is the hypotenuse. The "12" always represents the *base leg* of the triangle. The rise in inches per foot is the height leg or opposite leg. Thus, according to Pythagoras' proposition:

$$\text{Hypotenuse} = \sqrt{(\text{base})^2 + (\text{height})^2}$$

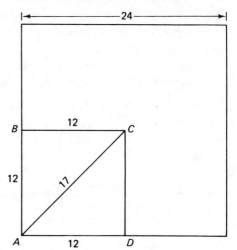

Figure 4–4 Fundamentals of
Rafter Layout

Let's apply this, using some easy calculations and Table 4–1 data. On Table 4–1 you can see that for a 9-inch rise per foot run the rafter length is 15 inches. Using the formula we find:

$$\text{Rafter length} = \sqrt{(12)^2 + (9)^2}$$
$$\text{per foot run} = \sqrt{144 + 81}$$
$$= \sqrt{225}$$
$$\text{per foot run} = 15''$$

You could try numerous other calculations, but they would all check out very closely.

Because a hip or valley rafter is also a right triangle, it too has a base leg and height leg which are used to determine the hypotenuse or rafter length. In trade usage the base leg has a proportionality of 17 inches per foot run. This is a slight error, as the Pythagorean theorem shows. To arrive at the figure (actually just under 17 inches) consider the square ABCD in Figure 4–4. Each of the four sides is 12 inches long and the square is on a single plane. Applying the formula

$$AC = \sqrt{(AB)^2 + (BC)^2}$$
and substituting $$AC = \sqrt{(12)^2 + (12)^2}$$
$$= \sqrt{144 + 144}$$
$$= \sqrt{288}$$
$$= 16.9706''$$

AC = diagonal
AB = side one
BC = side two
AB, BD and AC form
a right triangle

The reason 17 inches is used on the base leg is that a less than $3/100$ in error per foot run is induced in the layout. Let's prove the rounded-off 17-inch figure works.

From a framing square table of hip rafters for a 12-inch rise per foot run, we read $20^{8}/_{100}$ in length. Assume a run of 14 feet, calculate the rafter length per foot run:

Base = 17", and a rise/ft run 12"

increment
per ft run = $\sqrt{(17)^2 + (12)^2}$

= $\sqrt{289 + 144}$

= $\sqrt{433}$

incremental
increase/ft run = $\begin{array}{c} 20.8008" \\ \text{or} \\ 20^{8}/_{100}" \text{per ft run} \end{array}$

On the 14-ft run the rafter length would be 291.2" or $24^{26}/_{100}$ ft = 24 ft $3^{12}/_{100}$ in.

Jack Rafters must be cut and installed to complete a hip or valley section. There are two usual layout methods. You can use the tables on the framing square for either.

A jack rafter is cut so that one end has a common rafter layout and the other end has a compound miter layout. The compound mitered end is nailed to the valley or hip rafter.

There are two rafter layout methods: for the shortest jack rafter and the longest jack rafter. The shortest-jack rafter method is indicated on the face of the framing square. Its length and angle cuts are both given for a variety of on-center spacings, such as 16" and 24" oc. The length of the initial jack rafter is the incremental length added each time. For example, if the first jack measures 26½ inches, the second will measure 53 inches, the third 79½ inches, and so on.

If you use the longest-jack rafter method, you can calculate the incremental difference by dividing the number of the spaces between rafters by the total length of the longest rafter. This resulting length is the amount you subtract from each succeeding jack rafter. For example, assuming that the longest jack rafter measures 12'0" and that there are 5 spaces after the rafter layout on the plate (12 ft − 5sp = 28⅔ in.), you divide to establish the difference of 28⅔" per jack rafter. The second longest jack rafter would, therefore, be 115.2" long or 9'7.75" long.

Study these explanations carefully since they provide data for you to use in estimating materials and establishing training objectives. They will also help you to understand why rafters are laid out in prescribed patterns. You can also see that building techniques are based on sound mathematics.

Flat Roof

Really an extension of ceiling joists the flat roof is the simplest to build. (Review the earlier discussion in Chapter 3.) More accurately, joists serve as roof rafters, except that they are given a gradual slope to afford drainage. If wooden rafters are used, select the size and timber species from the tables in Appendices A and B. Factors that must be considered in their specifications include the Modulus of Elasticity or "E"; the extreme fiber in bending, or "F_b", live and dead loads, and on-center spacing.

Requirements for bridging are similar to those for the floor joists; and, as with joist construction, the crown on each rafter must be placed *up*.

If the rafter ends of a flat roof are inserted into a cement or block wall their ends must be tapered or beveled as shown in Figure 4–5. The reason for this is that it prevents the wall joint above the rafter from cracking as the rafter settles. The detail in the circle in Figure 4–5 shows what happens if the rafter ends are not beveled in this way.

As you can see, the cut must be tapered so that the top corner of the end bevel on the rafter meets at the inner edge of the wall.

Figure 4–5 Tapered End of Flat Roof Rafter

Truss Roof

There are two general types of prefabricated trusses which may be used on light commercial structures and townhouses: the monoplaner (single element thickness) and the ring-truss. Ordinarily the monoplaner truss is used.

The monoplaner truss is manufactured in various styles. Figure 4–6 shows five: the *pitched, Belgian, Pratt, flat,* and *scissors.*

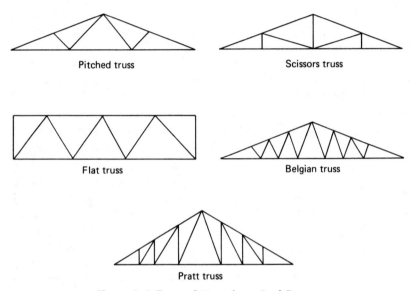

Pitched truss   Scissors truss

Flat truss   Belgian truss

Pratt truss

Figure 4–6 Types of Monoplaner Roof Trusses

Chords, web members, and *connector plates* characterize the monoplaner truss. The chords are what would normally be called rafter and joist in on-site roof building, and the web members are *usually* 2 × 4's inserted between the chords. They are held in place with connecting plates.

Figure 4–7 illustrates the pitched truss. Notice the various markings for the angle of web members and the maximum span for the 2 × 4 of the bottom chord. Let's examine this data further with the aid of Table 4–2.

The letter "X" represents the rise in inches per foot run. That means this specification is adequate for roofs with up to 6-inch rise (or pitch) per foot run. You can also see that the maximum length or span for a 2 × 4 chord truss is limited to a range of from 22 feet to 28 feet;

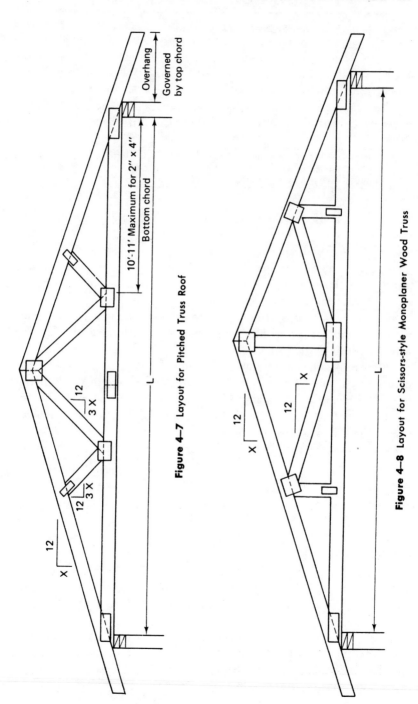

Figure 4—7 Layout for Pitched Truss Roof

Figure 4—8 Layout for Scissors-style Monoplaner Wood Truss

TABLE 4–2:   MAXIMUM SPANS FOR MONOPLANER 4 PANEL TRUSSES

X (pitch)	Length (in ft) 2 x 4's	Length (in ft) 2 x 6's
1-1/2	22	30
2	25	32
2-1/2	26	33
3	27	37
4, 5, 6	28	40

Note: Maximum spans for 2 x 4 and 2 x 6 top chords for 4-panel trusses are figured for
f = 1500 psi lumber with total loading of 55 psf including wind of 47.5 psf including
snow. All web members are made of 2 x 4's.

but if 2×6 chords are used, spans up to 40 feet may be covered with
the truss. These limitations are self-explanatory. Continue with the
use of "X" and you can see that where the web members are laid in
place, the angle is equal to 3 times for every 12 inches. An example
translates it thus: for a roof with a 5-inch rise, the layout is 15 inches
on the horizontal for each 12-inch rise.

This type of roof is assigned the following structural requirements
for grade and species of lumber F_b = 1500 psi, total load including
wind 55 psf, including snow 47.5 psf.

Figure 4–8 illustrates the configuration of a scissors-style monoplaner wood truss. Note that three vertical members distinguish this
truss. Its bracing webs are aligned at the same rise as the roof.

Both the trusses examined above may also be called *four-panel*
trusses. The panels are counted along the top chord. Two six-panel
trusses are shown in Figure 4–9. Since they are stronger, of course,

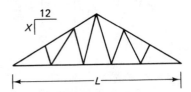

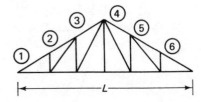

Figure 4–9 Two Types of Six-Panel
Trusses

they can be used over longer spans. The figures tabulated below show the allowable lengths or span for a given rise (✕ = inches of rise per foot run).

✕	Length
2½	46 ft
3	50 ft
3½	59 ft
4, 5, 6	60 ft

(Approximate maximum span with top and bottom chords of 2 × 6, F_b = 1500 psi, suggested pine or fir species.)

Truss Installation

When monoplaner trusses are delivered to the job site, they are usually bundled. Often their alignment is simplified if they are installed as they are taken from the bundles, even though they are designed to be equal (the left half is a mirror of the right half). Each truss must be hoisted to the wall plates and aligned with the wall studs. This is a job for several carpenters. Allowance must be made for overhang requirements before nailing them in position. Bracing is a must; tying-in must also be done before the sheathing.

Customarily, a 1 × 6 is tacknailed to either side of the ridge to keep the trusses in alignment. This means that a workman must walk on the lower chord. When it is practical, after three or four trusses are installed, a length of 2 × 8 planking should be laid across the chords as a walkway. Then a workman can easily advance the plank as more trusses are raised in place; and he always has a safe place on which to stand as he performs his tasks.

Ring Truss

Any style of truss may be manufactured with rings. It only means that the chords and web members are joined together with rings, and the members and rings are held in place with bolts. See Figure 4-10.

A circular dado is drilled in each place where a ring is to be installed. One-half the width of the ring fits into each dado, and a bolt draws the members tight. Ring trusses are extremely strong because of their design and because larger stock material (2 × 6, 2 × 8, 2 × 10, and 2 × 12 chords) can be used. They are usually employed

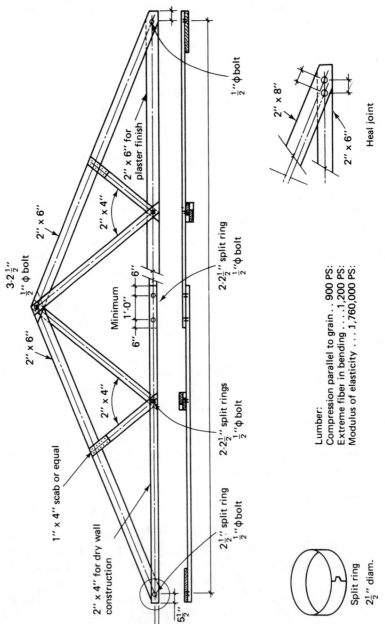

2" x 6" for
plaster finish

$\frac{1}{2}$" φ bolt

2" x 6"

3-2$\frac{1}{2}$"
$\frac{1}{2}$" φ bolt

2" x 4"

2" x 6"

2" x 4"

Minimum

6"

1'-0"

6"

2-2$\frac{1}{2}$" split ring
$\frac{1}{2}$" φ bolt

2-2$\frac{1}{2}$" split rings
$\frac{1}{2}$" φ bolt

1" x 4" scab or equal

2" x 4" for dry wall
construction

2$\frac{1}{2}$" split ring
$\frac{1}{2}$" φ bolt

5$\frac{1}{2}$"

2" x 8"

2" x 6"

Heal joint

Split ring
2$\frac{1}{2}$" diam.

Lumber:
Compression parallel to grain . . 900 PS:
Extreme fiber in bending . . . .1,200 PS:
Modulus of elasticity . . . . 1,760,000 PS:

Figure 4–10 Layout for Ring Truss Roof

with roofs requiring slopes up to 7 inches and with greater than 24″ oc spacing.

Installation: If the truss is to sit in or rests against piers or pilasters, anchor plates are installed on those members. Then the truss is usually bolted to the plate.

Because of the heavy weight of these larger trusses and because of the extended on-center separation of the designs, there is more danger during installation. Workmen are required to brace each truss diagonally from its top chord to the bottom chord of the previously installed truss. Extension ladders and cranes with work buckets are frequently used to position the men for work. Remember, these and similar devices can be dangerous so great care must be used during their installation.

Summary

Whether built on-site or pre-fabricated, the truss roof is an excellent roof for many commercial structures and townhouses since it eliminates the need for interior load-bearing partitions. It requires prior study, however, and the careful formulation of an erection plan. The installation of a monoplaner truss presents some danger to workmen but it can be minimized with the use of walk-way planking and proper bracing. Installation of a ring truss roof not only presents different working conditions but also additional risks which often can not be completely eliminated.

Rafter Positioning and Reinforcement

Some reinforcement is required wherever truss roofs are installed. This is usually done with angle braces from ridge line to plate line and with gable-end studding.

In addition to angle braces and gable-end studding the bracing on "stick built roofs" must include *collar ties* or *collar beams*. One-inch stock, usually beveled at both ends, is nailed approximately 18 inches down from the ridge and parallel with the joists or plates. These ties distribute the downward force and relieve the pressure that is being exerted on the outer walls.

Much has been said about rafter framing; however, one point again needs to be emphasized. If the MOD 24 System is used the rafter layout must conform to that 24″ stud spacing and stud placement. Unified structural strength can only be maintained if these requirements are met.

ROOF SHEATHING AND COVERINGS

Plywood is the principal sheathing material in use today regardless of the method of roof construction. There are exceptions, of course, where insulated fire-retardant panels are installed. In such cases, follow the manufacturer's installation specifications.

Figure 4–11 shows plywood layouts for two typical commercial applications. The design at the left shows flat decking with built-up roofing and ¾" sheets laid over joists spaced 48" oc. Note also, the panels require either edge blocking or must be tongue and grooved at the edges. The design at the right for a sloping roof, in contrast, has the sheathing installed with plyclips as a reinforcing device. T&G plywood is sometimes specified for this sort of roof too.

Table 4–3 provides data specifically calculated for plywood that is to be used as roof decking. In the panel identification index expressed in the left-hand column, the figures left of the slash mark (/) represent the on-center inch separation of rafters. The 2-4-1, 1⅛" and 1¼" index numbers identify the very thick plywood panels that can be used where spacings of rafters, joists, or trusses increase up to 72" oc. If you study the ten columns at the right, you will see that the allowable live loads (psf) drop off rapidly when the on-center spacings are 60 inches and wider, even though the panels are very thick.

Installation

Where conventional on-center spacing of rafters is maintained, the method of installation will generally follow residential techniques, especially on sloping roofs. This means that you must use installation techniques such as breaking joint with successive rows, ensuring that a plywood sheet covers two or more spans, and including proper reinforcement between spans.

Application of these techniques is dictated by the design factors of the roof and the sheathing selection. For example, you can see from Table 4–3 that the allowable live load would be 50 psf if ⅜" sheathing were used on a roof with rafters spaced at 24" oc.

What the table does not tell you is that if the edge is unsupported, both the shingle installation and the roof maintenance will be greatly impaired. Application must therefore include edge supports. The least expensive, adequate reinforcement for most roofs that carry a limited live and dead load is the plyclip. In addition to reinforcement, it provides the needed spacing of plywood sheathing to allow room for swelling.

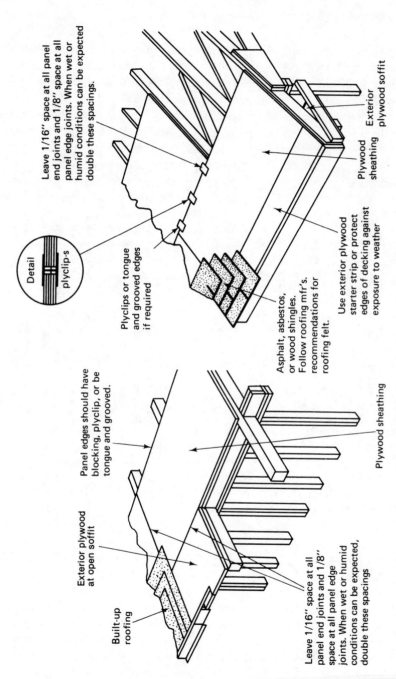

Leave 1/16" space at all panel end joints and 1/8" space at all panel edge joints. When wet or humid conditions can be expected, double these spacings.

Exterior plywood soffit

Plywood sheathing

Detail

plyclip-s

Plyclips or tongue and grooved edges if required

Asphalt, asbestos, or wood shingles. Follow roofing mfr's. recommendations for roofing felt.

Use exterior plywood starter strip or protect edges of decking against exposure to weather

Panel edges should have blocking, plyclip, or be tongue and grooved.

Plywood sheathing

Exterior plywood at open soffit

Built-up roofing

Leave 1/16" space at all panel end joints and 1/8" space at all panel edge joints. When wet or humid conditions can be expected, double these spacings

Figure 4–11 Layouts for Plywood Roof Sheathing

TABLE 4-3:   SPECIFICATIONS FOR PLYWOOD ROOF DECKING

Panel ident. index	Plywood thickness	Max. span (inches)	Unsupported edge-max. length[d]	Allowable live loads (psf) Spacing of supports center to center (inches)									
				12	16	20	24	30	32	36	42	48	60
12/0	5/16"	12"	12	150									
16/0	5/16", 3/8"	16"	16	160	75								
20/0	5/16", 3/8"	20"	20	190	105	60							
24/0	3/8", 1/2"	24"	24	250	140	95	50						
32/16	1/2", 5/8"	32"	28	385	215	150	50	40					
42/20	5/8", 3/4", 7/8"	42"	32		330	230	145	90	75	50	35		
48/24	3/4", 7/8"	48"	36			300	190	120	105	65	45	35	
2-4-1	1-1/8"	72"	48				390	245	215	135	100	75	45
1-1/8" Grp. 1 & 2	1-1/8"	72"	48				305	195	170	105	75	55	35
1-1/4"	1-1/4"	72"	48				355	225	195	125	90	65	40

Notes:

a. These values apply for C-D INT-APA, C-C EXT-APA, STRUCTURAL I and STRUCTURAL II C-D INT-APA, and STRUCTURAL I and STRUCTURAL II C-C EXT-APA grades. Plywood continuous over 2 or more spans; grain of f face plies across supports.

b. Use 6 d common smooth, ring-shank or spiral-thread nails for 1/2" thick or less and 8d common smooth, ring-shank, or spiral-thread for plywood 1" thick or less. Use 8d ring-shank or spiral-thread or 10d common smooth-shank nails for 2.4.1, 1-1/8" and 1-1/4" panels. Space nails 6" at panel edges and 12" at intermediate supports, except that where spans are 48" or more, nails shall be 6" at all supports. Space panel ends 1/16", and panel edges 1/8". Where wet or humid conditions prevail, double these spacings.

c. Special conditions, such as heavy concentrated loads, may require constructions in excess of these minimums.

d. Provide adequate blocking, tongue and grooved edges or other suitable edge support such as Plyclips when spans exceed indicated value. Use two Plyclips for 48" or greater spans and one for lesser spans.

e. Uniform load deflection limit: 1/180th span under live load plus dead load, 1/240th under live load only.

Customarily sheets of ½" thickness or less are nailed with 6d common smooth, ring shank, or spiral thread nails. Fasten plywood panels over ½" thick to 1" thick with 8d nails; and for those over 1" thick, use 10d nails. Space nails 6" at panel edges and 12" at intermediate supports, except where rafters are spaced 48" or greater. Then space 6" on all supports.

Roof Coverings

Two types are used: the strip shingle and the built-up roof. Either or both may be subcontracted, although strip shingles are often installed by carpenters. Figure 4–11 on page 117 shows each type.

The techniques for installing built-up roofs need further examination. For example, the specifications might stipulate that the roof is to be covered so as to qualify as a 20-, 30-, 40-, or 50-year roof. So various grades and weights of tar paper (felt paper) will have to be used, since combination of grade and weight, number of layers, and overlap are the factors that determine the life cycle of the roof covering. The roof consists of 2 to 4 layers of *base sheets* of 15-pound felt, topped or surfaced. The per-square *surface material* typically consists of 300 to 400 pounds of stone aggregate held in place by 50 pounds of hot asphalt.

Installation of Built-up Roofing

Each layer of built-up roofing underneath the surface is called *a ply* (Figure 4–12). In a 5-ply roof, for instance, the first two layers are laid without binder (melted pitch or asphalt) and are called *dry nailers*. The remaining three layers are installed with the binder underneath the felt paper. The exposure is usually set for 10 inches to provide the five plys.

The installation begins at the eaves of the roof so that the strips overlap in the direction of the water shed. A series of three to five rows of felt paper are laid in position. Then the mop man wipes a layer of pitch or asphalt ahead of the roll; and the broom man follows along, brushing the felt paper into the hot pitch or asphalt. This binds the two products and eliminates air pockets at the same time.

On sloping roofs the felt may be nailed along the top edge of the paper to hold it in position. The length of felt is folded back (but not creased) and pitch is mopped on the roof surface; and, as before, the broom man unrolls the felt and brushes it into the pitch in order to make the two bond.

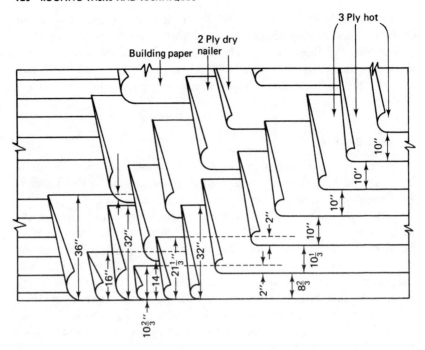

Figure 4–12 Layout for 5 Ply Built-Up Roof

Overhangs

A distinctive feature of many contemporary townhouses, office buildings, and stores is their use of roof surfaces which are almost perpendicular. The shingling on the roofs of such townhouses is usually applied so that it covers the upper wall of the two-story building. Its projection over the first-floor exterior wall is usually not more than 12 to 15 inches. Figure 4–13 shows an example of an overhang roof-wall finished in this manner. The framing required can either be similar to studding or it may incorporate a partial truss assembly whose members form an outer beveled surface and an inner wall stud. The ceiling joists are then also the flat roof rafters.

For the store front whose designs call for projections of 48 inches or more, a modified saw-tooth truss is usually built. Figure 4-14 shows both the basic saw-tooth truss and the modified truss for this application.

Simple to fabricate and simple to install, this one truss unit establishes three surfaces for exterior finishes. The bottom is the soffit; the face functions as wall surface or fascia, and the top is a roof surface.

Figure 4–13 Building with Part of Roof Serving also as Upper Floor's Wall Surface

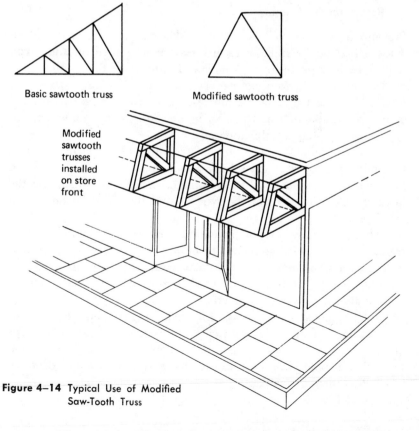

Basic sawtooth truss

Modified sawtooth truss

Modified
sawtooth
trusses
installed
on store
front

Figure 4–14 Typical Use of Modified
Saw-Tooth Truss

With the emphasis on quality, accuracy, and the use of secure fastening techniques, the overhangs require specialized construction. All aspects of this roof construction, like the others, require a major emphasis on training to assure a quality product.

TRAINING ON ROOFING

The roof construction skills that carpenters need are varied. They range from the simple erection of truss assemblies through formal layout to the cutting and installation of complex hip, valley, jack, and cripple rafters. While the work crew for the job is being hired, a comprehensive question and answer period should be made a part of each interview. With respect to roofing, you should tailor the questions to the roofing principles that will be employed. Then be sure to fill out the training record as each interview is conducted. After that, you can develop the training plan.

Recording the Training

Previous training as well as degree of skill and knowledge should be made a part of the training record. Carefully worded questions will require each carpenter to respond specifically rather than with a "yes" or "I've done that before" answer. You need to know how many times or how long he has done certain tasks and whether he was trained only in the trial-and-error school or in a classroom or in an apprentice program, or in some combination of these.

A crew's training record for roofing skills and knowledges should be similar to that in Figure 4–15, although some entries would not be applicable to your job. List only the ones necessary to support the tasks and skills needed for the type of roof that is to be built on a given project.

After you study the completed training record, you can make a training plan. Where skills have to be learned, you must try to integrate training periods in the regular work schedule. At first, the response may be "no training for me." Or the contractor may even say "I can't afford the loss in production."

Both responses are wrong. Effective sessions can be only 30-minutes long if they are efficiently prepared and presented. For instance, they may be scheduled just before noon or before quitting time since people have a natural tendency to slack off and decrease production then, in any case.

Name of employee	Skill										Remarks
	Common roof layout	Rafter layout	Erection	Hip and valley roof layout	Hip, valley and jack rafter cutout	Hip and valley erection	Truss roof construction and erection	Sheathing	Shingling		

◪ Basic ⊠ Skilled Ⓢ Specialist

◩ Some ☐ None Ⓐ Apprentice

Figure 4–15 Training Record of Roofing Crew

Finally, not all personnel need to be trained in all aspects of a given job. Teach all the unqualified personnel the principles of erection, bracing, and other assembly phases; then select one or two who are nearly qualified to learn to do the specialized work.

Task Activities

To aid both trainer and trainee the following group of task activities details the variety of knowledge and outlines the skills necessary for roof construction. You will find the basics included as well as the complex tasks of rafter layout, a variety of assembly sequences, and many details about roof sheathing and covering.

RF 1 TASK ACTIVITY: PREPARING RAFTERS

RESOURCES

Estimated Manhours for 1 man: 20–30 mins for first rafter
2–3 mins each add'l rafter

Materials:

roof layout plans
rafter lumber stock

Tools:

50-ft tape measure	framing square
crosscut saw	portable power saw
combination square	

PROCEDURES

A. COMMON RAFTERS:
   1. Calculate the rafter run from data contained in the building plans; also determine the pitch of the roof. Then use Table 4–1 to calculate the total length of the rafter.
   2. Lay a rafter across a pair of sawhorses and find its crown edge. Using the framing square, mark a ridge cut at one end of the stock with longest point at the crown.

3. Using your tape measure, place the tape from the long point on the ridge cut toward the opposite end of the stock and mark the rafter length. If an overhang is needed add its length to the mark just made.

4. From the first mark, lay out bird's mouth for the rafter. From the second mark, trace the overhang vertical and horizontal cuts. Use the crosscut or power saw to cut away excess stock.

5. Before cutting the ridge, allow for one-half of the thickness of the ridge stock as trim from original layout.

B. *HIP OR VALLEY RAFTERS:*

1. Position rafter stock across a pair of sawhorses. Find and mark the crown edge; then mark a ridge cut using the rise per foot and 17" (12" is used on common rafters.)

2. Calculate the rafter length from rafter table on the framing square or use the step-off process. Make another mark at rise per foot and 17" and complete the bird's mouth.

3. Calculate and lay out the rafter overhang.

4. Using the combination square, mark 45 degrees at the ridge and end of the overhang so that when cut the longest point (or edge) is even with the first layout marks on the stock and is one-half the thickness of the rafter or ¾".

5. Set power saw for 45-degree angle and cut ridge and overhang. Then use hand saw and power saw to trim the bird's mouth.

C. *JACK RAFTERS:*

1. Prepare one end of a piece of rafter stock with a common rafter layout. On a hip this will be the bird's mouth layout. On a valley, a ridge cut. Make all marks so that the crown is up.

2. Calculate jack rafter length by using the framing square tables and/or tape measure on the roof. Lay out the opposite end of the rafter, using the framing square and members indicated on its scales. Then, using the combination square, mark the end to fit the hip or valley for a 45-degree cut. (This cut is made from one side through the entire stock.)

3. Measure and lay out overhang to required length. Make proper vertical and horizontal layouts (hip jacks only).

4. Cut rafter according to layout marks. Measure, mark, and cut a second and third jack rafter; make each one longer than the first by adding the amount indicated on the framing square.

RF 2 TASK ACTIVITY: PREPARING A MONOPLANER TRUSS

RESOURCES

Estimated Manhours for 2 men: 4–8 hrs for fabrication of truss jig
1 hr (avg.) for truss assembly

Materials:

building plans	web stock
truss plans	chord stock
connecting plates	truss jig materials

Tools:

tape measure	bevel square
hammer	power saw
chalk line	crosscut saw
framing square	

PROCEDURES

1. Fabricate a jig from 2 X 4's approximately 26 inches above ground level. Space horizontal 2 X 4 members where web and chords will join. Snap chalk lines along the outer limits of the chords and fasten block of 2 X 4 to the outside of these lines.

2. Cut and place the chords in the jig. Mark the position of each web; lay out and cut webs; put webs in place and use blocks of 2 X 4 to hold alignment.

3. Pre-cut at the saw the required number of chords and webs. Assemble by using connector plates.

RF 3 TASK ACTIVITY: ROOF RAFTER ERECTION

RESOURCES

Estimated Manhours for 2 men: 2 hrs (avg.) for layout
15 mins (avg.) per rafter erection
(includes bracing)
2 hrs (avg.) gable end construction

Materials:

pre-cut rafters
bracing stock, 2 × 4's & 1 × 8's
ridge stock
building plans

Tools:

ladders
hammer
plumb bob
crosscut saw

scaffolds as required
level
framing square

PROCEDURES

1. Lay out rafter placement on plates so that each rafter rests over a stud. Measure, mark, and cut a corresponding ridge member; make its layout equal to the plate layout.
2. Position rafters along the walls with ridge cut toward the sky. Lay some plywood walking surfaces through the center of the building on top of the ceiling joists.
3. Nail two rafters to the ridge, one near or at the end and one 6 to 8 ft away. Raise the ridge so that the bird's mouth end of the rafters seats against the plate; position and nail.
4. Using 2 × 4 stock, support the ridge temporarily. From the opposite side of the building install the matching rafter, first nailing at the bird's mouth then at the ridge.
5. Install several more sets of rafters; then install temporary angle brace from ridge to outer plate or across several ceiling joists. When installing brace, use plumb bob to bring the ridge into alignment.
6. Install all rafters. Then install collar ties across every other or every third set of rafters. Cut and install permanent diagonal braces at gable end.
7. Install gable end studs by aligning each stud over wall stud.

   NOTE: On hips and valleys, the common rafter roof sections are installed first, then the hips and valleys. Hip and valley jack rafters are installed last.

RF 4 TASK ACTIVITY: ROOF MONOPLANER TRUSS ERECTION

RESOURCES

Estimated Manhours for 2 or 3 men: 15 mins per truss (avg.)

Materials:

building plans
bracing stock
trusses

Tools:

hammers
plumb bob
ladders
level
line (heavy cotton or nylon cord)
2 × 8 or 2 × 10 planking
lifting machine (*optional*)

PROCEDURES

1. Lay out the truss placement on wall plates and over wall studs. Along one side wall install line even with the overhang distance of each truss plus ¾".

2. Position a truss (not the end one) on the plates ¾" back from the line and in line with the truss placement; nail the truss to the plates.

3. Nail a temporary 1 × 4 brace from the top chord to the end wall plate; make truss vertical by using plumb bob.

4. Position next truss properly and nail in place. Space the top chord by installing a temporary 1 × 4 across the top of the two chords; continue across the length of the building. Use 2 × 8 or 2 × 10 planking for walkways while installing trusses and spacer boards.

5. Install gable end trusses. Brace near the top of the gable end and down to an intermediate truss *after* first spacing the lower chords with a permanently nailed 2 × 4 spacer board.

6. Install plywood sheathing.

RF 5 TASK ACTIVITY: STORE FRONT ROOF OVERHANG

RESOURCES

Estimated Manhours for 2 men: 1 day for 8 lin ft

Materials:

truss sections
building plans
bracing
sheathing
felt paper and shingles or other covering

Tools:

hammers   crosscut saw
level   mason line
ladders   scaffold
portable power saw

PROCEDURES

1. Fabricate modified scissors truss sections for entire front; determine anchoring method and prepare anchoring surface.
2. Raise and install a truss section at each end of the front. Position a line between installed trusses so it stands off ¾" along the lower outer edge.
3. Install the remaining truss sections at desired on-center spacing. Ensure that the lower edges are perfectly in line.
4. Nail a 1 × 4 temporarily along the top of the trusses to align each truss for on-center spacing.
5. Cut and install sheathing plywood; nail first to vertical members then to top members. (Remove 1 × 4 before nailing the sheathing to top members.) Cover sheathing with 15-lb felt.
6. Sheath back areas of trusses if they are exposed; cover with felt.
7. Install shingles or other facing material on vertical portion of overhang. Apply additional base layers and surface material to flat and rear portions of the truss.

REVIEW QUESTIONS

1. Structural members having a deflection limitation of $\int/180$ may be used for some rafters. True or False?
2. Selecting a rafter by species and grade involves the defining of the rafter's _____ and _____.
3. What are the two methods of laying out a common rafter?
4. Explain why the "17 inch" is used on the framing square in lieu of the "12 inch" when stepping-off a hip rafter.
5. Does one end of a jack rafter have a compound miter cut? Define jack rafter.
6. What does using the "shortest jack rafter" method mean?
7. Why is it essential that the end of a flat roof rafter be tapered?
8. List the five types of monoplaner trusses.
9. Name and define each member of a monoplaner truss.
10. What is the maximum span for a 4-panel monoplaner truss?
11. A 6-panel monoplaner truss with a 5" rise may span a width of _____ feet.
12. What is a ring truss?
13. How does rafter placement contribute to the 24" system of framing?
14. What is the roof sheathing nailing-schedule for ⅜" and ½" thicknesses?
15. Define a typical built-up roof.
16. How can a partial truss be used to create the base for a commercial building overhang?

5

Exteriors of Commercial Structures

Composition: the underlayment, insulation, and finish materials that go into making either a wall system or an assembly

Cornice: the projecting horizontal trim that serves as finish along the roof line of a structure

Double-wall siding system: a construction method that uses both sheathing and siding as an exterior wall surface

Fascia: the relatively broad, flat, vertical surface of a cornice that is nailed across the rafter ends and onto a fly rafter

Finish carpenter: an experienced journeyman skilled in the fine finish work of carpentry

Flashing: a piece of metal molded to fit against a wall and over an outcropping or window cap; used to seal against water leaks

Glazier: a person who installs glass; particularly the glass and aluminum store fronts used in commercial construction

Insulation: the material used to prevent moisture and heat or cold from penetrating into a building

Joinery: craft or trade of constructing doors, windows, sashes, paneling and other permanent woodwork for interiors; also the highly skilled carpentry using special cuts and patterns to unite and join wood members, as for a drawer

Lookout: the 2 × 4 structural member nailed horizontally from the lower rafter edge overlay to the wall's outer surface

Quality: a characteristic with respect to excellence, as in skilled workmanship and in high grade of materials

Reliability: the quality or qualities and skill possessed by a workman who can be counted on to do what is required on the job; also the dependability of a system because of the excellence of its composition

Siding: any covering that is affixed to the outside or exterior walls of a structure; usually of overlapping boards or shingles, veneers, or wooden or metal paneling

Single-wall siding system: a construction method that uses only siding over the framing as an exterior wall surface

Soffit: the horizontal underside of an eave or cornice; also the stock used in a closed or boxed cornice extending from fascia to frieze

Commercial carpentry activities on the outside of a structure may be very limited if brick, stone, or concrete walls are used. On these buildings carpentry tasks would probably include sheathing walls, building cornices, and installing door units and store fronts. But at times there may be extensive use of carpentry. For example, if a townhouse is to be finished with T-1-11 or 303 rough sawed plywood, much time and many skills will need to be used. See Figure 5-1.

Except for sheathing installation, exterior carpentry is *finish* work. Classification of the work as "finish" is very important because it is significant in terms of quality workmanship and cost of materials. As a rule only highly skilled craftsmen are given the responsibility for installing the siding, cornice, and related materials. The work must be exacting and sound. If joints are made they must be properly fitted; if flashing is installed, it must be accurately positioned. Where building lines such as corners and cornices are developed, they must run true. Much prior training is needed, therefore, as well as dedication.

To do such exterior carpentry work, the characteristics of the products and the appropriate installation techniques must be well understood. They will be presented in this chapter with considerable data. Because of the great variety of materials available on today's market, representative examples have been chosen to illustrate groups of related products. It is up to you to transfer an installation technique for a product by associating it with similar available products. For example, there will be a discussion of the plywood paneling used for

Figure 5-1 Exterior Siding *(Courtesy of American Plywood Association)*

exteriors. If Type 303 panels with battens are called for in a plan you will then recognize that the technique for their installation would parallel techniques for T-1-11 or Masonite panels with battens.

ORGANIZATIONAL FUNCTIONS IN EXTERIOR CONSTRUCTION

Before working out the task activities in detail, the organizational function and total requirements for this phase of the construction job must be examined. The major subphases of the exterior work phase were mentioned earlier, including store front, sheathing, siding, cornice and door installations. All require considerable attention in the planning stage. As shown in Figure 5–2 several subcontractors may be employed to perform selected subphases of the exterior work. If they are, the contractor must evaluate what has to be done beforehand. In order to provide subcontractors with materials and method data and scheduling requirements the contractor must, therefore, develop this phase of the work thoroughly.

The data that subcontractors need will, of course, concern their specialties, and must include any restrictions and limitations. Providing them with a detailed schedule may take considerable time to develop, but it is time well spent since their work must be integrated with the total job.

In addition, the time-line plan for the main work force must be carefully made. A conference with staff members and foreman usually produces questions about scheduling, supply ordering, personnel training, and production rate. When all such questions are answered, each member of the team should be able to do his part as required. Moreover, one or the other of the subcontractors may also have questions which indicate that he needs more time for job completion than originally planned. That too will be a problem to resolve as early as possible.

To satisfy the material requirements called for in the plan certain matters must be examined by the person in charge of materiel. Will the materials be available when needed? Standard materials are usually ready for use in this work area; however, a thorough check must be made to ensure their availability and delivery. Long-lead items such as specialized facings, sidings, and trimmings may also be needed. In short, the materiel man needs to size up the whole picture.

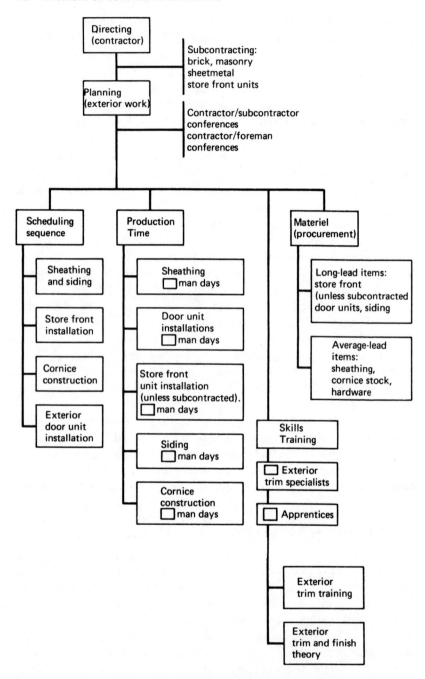

Figure 5-2 Organization of Exterior Construction Activities

When the production schedule is made, it will have to include the work subphases to be performed by each subcontractor as well as an evaluation of the crews' skills. It may well be possible to provide OJT, but there will probably have to be one or more experienced *finish carpenters* on the job.

In summary, this phase of the construction job is treated like the others. That is to say, its planning must answer the same six important questions:

1. Who will do the work?
2. How much will each crew or group do?
3. What materials will be needed?
4. How much time will be used?
5. What skills will be needed?
6. What training will be required?

The organization for exterior carpentry must function as a unit and any decisions made must be made a part of the time-line and training plans. If the plans look good, the materials, methods, and procedures which make up the details and descriptions of the various job activities must then be considered thoroughly.

EXTERIOR CONSTRUCTION TECHNIQUES

Following a brief examination of the variety of sidings and quantitative estimating, cornices will be studied. Then door units and store front installations will be reviewed. Let's first examine two siding systems in order to study their principles of application technique, relative cost, quality reliability, and any need for special preparations.

Single-Wall Siding Techniques

Figure 5–3 illustrates the single-wall siding technique used on a townhouse. The exterior of the units shown is Textured 303 plywood siding grooved 8" oc. This system is distinguished by the fact that it is applied *directly* to the framing. Direct application then is the key to what is called the single-wall siding system.

Single-wall composition means that the system consists of a single panel of exterior siding installed vertically or horizontally directly onto studs. Figure 5–4 illustrates both vertical and horizontal installation techniques. You will notice that there is *no* sheathing or building paper

Figure 5–3 Single-Wall Siding System *(Courtesy of American Plywood Association)*

installed between stud and siding. You will also note that corner bracing is not required.

But composition specifications of the wall prescribe $\frac{7}{16}''$ or thicker panels wherever 24″ framing systems are used. If a conventional 16″ oc framing system is used, panels as thin as $\frac{5}{16}''$ may be used. Nailing at 6″ oc along the edges and 12″ oc on intermediate studs develops a structural rigidity that exceeds minimum specifications in most locales.

Cost Factors: The savings are easily defined by the reduced requirements for material and labor. Materials savings result because sheathing, building paper, and corner bracing materials are not used. Labor savings result because installation times are not required, of course, for the materials so eliminated. This means that, if given a choice of wall siding systems, the contractor's bid can be very competitive if he selects this single-wall system.

Reliability and Quality: The application of a panel of siding 4 × 8 (or larger) to a standard stud wall gives that wall a panelized structural quality. Panelized structures are very strong and resist pressures from numerous angles in a variety of velocities. When built properly they withstand great force. Boxes and many other products manufac-

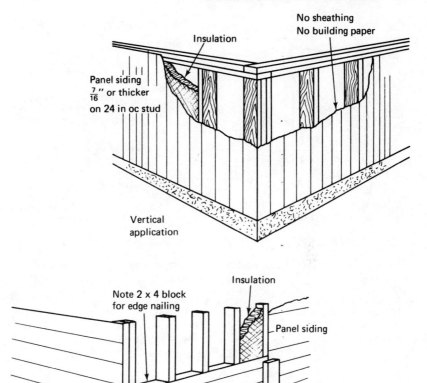

No sheathing
No building paper

Insulation

Panel siding
$\frac{7}{16}$ " or thicker
on 24 in oc stud

Vertical
application

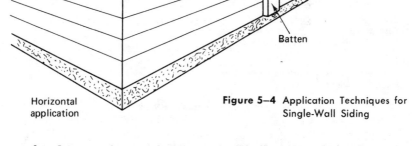

Note 2 x 4 block
for edge nailing

Insulation

Panel siding

Batten

Horizontal
application

Figure 5–4 Application Techniques for
Single-Wall Siding

tured today use the panelized system. The concrete form discussed in Chapter 2 is also an example of a panelized structure.

To illustrate one more factor of reliability and quality consider the data on insulation provided by the American Plywood Association in Figure 5–5. Ratings in the left diagram show that properly insulated single-wall siding affords the same reliability as a double-wall and meets or exceeds FHA-MPS insulation requirements. The right diagram in the figure shows data for a double-wall siding system whose insulating quality is no better.

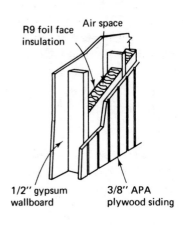

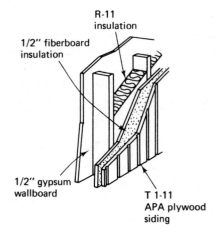

Single wall*	Resistance**	Double wall*	Resistance**
Outside air	= 0.17	Outside air	= 0.17
3/8" Plywood	= 0.47	5/8" Plywood	= 0.78
R9 Insulation	= 9.00	1/2" Fiberboard sheathing	= 1.32
3/4" Minimum air space	= 3.48	R11 Insulation	= 11.00
1/2" Gypsum board	= 0.45	1/2" Gypsum wallboard	= 0.45
Inside air	= 0.07	Inside air	= 0.68
	R = 14.25		R = 14.40
	U = 1/R = 0.07		U = 1/R = 0.07

‡R13 insulation (3-5/8") with or without
foil may be substituted.

* Meets or exceeds
FHA-MPS insulation requirements
for multifamily construction, all regions.

**R values from ASHRAE Guide, Actual
thickness and species groups affect plywood
insulation values.

Figure 5–5 Insulation Factors in Wall Systems

In summary, panel siding used on framing affords good rigidity and resistance and insulation reliability and quality.

Special Preparations for Installation: The nailing schedule has already been defined. Several other details related to joint installations need to be defined. Figure 5–6 illustrates the techniques used to join panels in most places where stress and fit are a concern. For instance, one extremely important idea expressed in each diagram is *solid blocking.* Notice especially the horizontal wall joints, *butt and flush.* Wherever panels are installed horizontally or wherever the wall section height exceeds the plywood length, the flashing is installed on solid backing stock. This application is easily made and eliminates both leaks to the inner wall, and the need to overlap panels. That is especially important wherever character lines must be maintained.

Now let's examine the double-wall siding technique from the same points of view as for single-wall siding.

Double-Wall Siding Technique

Unlike single-wall siding which is nailed directly to the studs, double-wall siding is nailed to the wall sheathing in the double-wall siding technique. The obvious gain from using a double-wall siding system is the ability to use a variety of siding materials which, in themselves, may not have the stiffening reliability needed.

Composition of the Wall: The Wall consists of the sheathing material and siding. Figure 5–7 shows an example of a double-wall siding system—in this case lap siding. In double-wall systems the sheathing materials may be fiberboard made from fibers and asphalt or plywood of various grades and thicknesses. Some are tabulated below:

Sheathing Panels	Minimum Thickness	Maximum Stud Spacing
12/0, 16/0, 20/0	$\frac{5}{16}''$	16''
16/0, 20/0, 24/0	$\frac{3}{8}''$	24''
24/0, 30/12, 32/16	$\frac{1}{2}''$	24''

Standard 1×6, or 1×8 S4S or T&G pine or fir lumber may also be used as sheathing.

If fiberboard and some grades of plywood sheathing are used, building paper or breather-type paper need not be used. If boards are used for sheathing, however, the paper is recommended.

On the outer surface of the sheathing all varieties of siding may be installed. It should be nailed through the sheathing to the studs whenever possible.

Cost Factors: The cost of completing exteriors with double-wall siding systems is generally greater than for single-wall systems. There are several cost factors which can be modified however, so as to affect savings. First, use a wood or fiber sheathing board that eliminates all need for either let-in or diagonal 2×4 bracing. Remember, when you select the sheathing, always choose materials that will hold the siding

VERTICAL WALL JOINTS

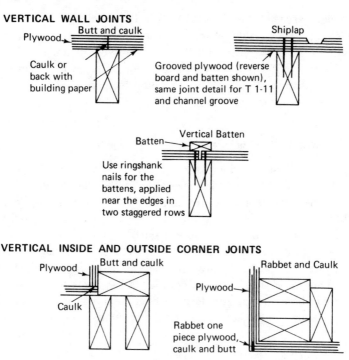

Butt and caulk

Plywood

Caulk or
back with
building paper

Shiplap

Grooved plywood (reverse
board and batten shown),
same joint detail for T 1-11
and channel groove

Vertical Batten

Batten

Use ringshank
nails for the
battens, applied
near the edges in
two staggered rows

VERTICAL INSIDE AND OUTSIDE CORNER JOINTS

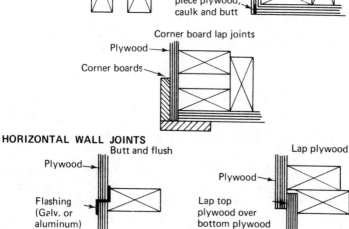

Butt and caulk

Plywood

Caulk

Rabbet and Caulk

Plywood

Rabbet one
piece plywood,
caulk and butt

Corner board lap joints

Plywood

Corner boards

HORIZONTAL WALL JOINTS

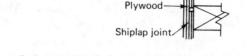

Butt and flush

Plywood

Flashing
(Galv. or
aluminum)

Lap plywood

Plywood

Lap top
plywood over
bottom plywood

Shiplap

Plywood

Shiplap joint

Figure 5–6 Joint Details for Plywood Siding *(Courtesy of American Plywood Association)*

HORIZONTAL BELTLINE JOINTS

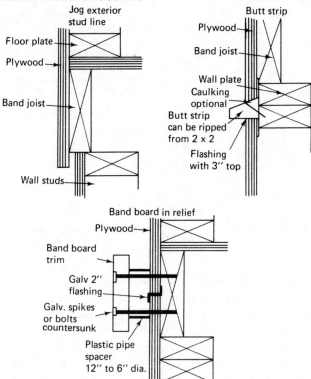

Jog exterior stud line

Floor plate

Plywood

Band joist

Wall studs

Butt strip

Plywood

Band joist

Wall plate

Caulking optional

Butt strip can be ripped from 2 x 2

Flashing with 3" top

Band board in relief

Plywood

Band board trim

Galv 2" flashing

Galv. spikes or bolts countersunk

Plastic pipe spacer 12" to 6" dia.

WINDOW DETAILS

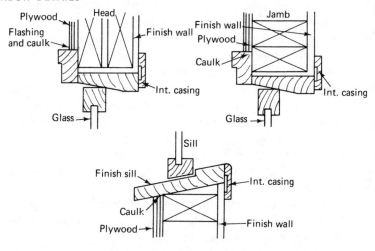

Plywood

Head

Flashing and caulk

Finish wall

Int. casing

Glass

Finish wall

Jamb

Plywood

Caulk

Int. casing

Glass

Sill

Finish sill

Int. casing

Caulk

Finish wall

Plywood

Figure 5–6 (continued)

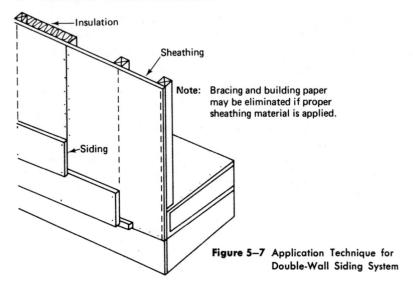

Figure 5–7 Application Technique for Double-Wall Siding System

nails securely. All manufacturers have such data. For instance, if $\frac{5}{16}''$ or $\frac{3}{8}''$ plywood sheathing is used, lap siding must be installed with either 6d or 8d nails that penetrate the studs behind the sheathing. There are also fiberboards so designed that the sidings nailed to them with special *ringed* or *barbed* nails will hold well.

Ordering the siding pre-cut to exact lengths can result in two more savings. One, because mass production cutting at the saw to specific uniform lengths eliminates most jointing and reduces time at the saw. Two, having siding in specified lengths greatly reduces or minimizes wasted footage. If you plan with these factors in mind, be sure to tell the foreman. He must know and he must advise those on the crew who are siding specialists exactly how the siding will deliver.

Finally, within a given kind of siding there may be various qualities or grades. Study the job specifications carefully. Select the grade that meets the specifications. But keep in mind that a grade only minimally acceptable might require additional preparatory work. That also must be added to job costs. For example, if the contract specifies that a surface must be painted and guaranteed for three years, a poor or minimal grade of siding may require one or more of the following: 1) added time for installation because of splits, checks, or breakage; 2) added preparatory steps such as sealing knots, caulking, and filling splits; and 3) extra coats of paint to seal and protect the surface to meet the three-year commitment.

A cost analysis study must be part of the selection process for the sheathing and siding that go into a double-wall siding system.

Requirements: We touched a bit on requirements previously. Unless otherwise substantiated by reliable manufacturers' claims, the general rules for application of *sheathing* are:

1. Where $\frac{5}{16}''$ panels are installed, stud on-center spacings can not exceed 16 inches.
2. Where $\frac{3}{8}''$ or thicker panels are installed, stud on-center spacing may be up to 24 inches.
3. That nailing will conform to standards where 6d nails are used for $\frac{5}{16}''$ and $\frac{3}{8}''$ sheathing and 8d and 10d nails are used for sheathing thicker than $\frac{3}{8}$ inches.

The general rules for application of *siding* are:

1. When used, nails must penetrate into studding.
2. Joints shall be properly backed and butted in accordance with swelling and shrinking requirements as stipulated by the manufacturer.
3. Caulking, flashing and other weatherproofing shall be used to the extent required to do the job intended and not to cover shoddy workmanship.

Reliability is a significant overall factor in planning for a double-wall siding system, even though its contributing elements establish the points of reliability. Your reliability guarantee may have to detail the structural values of the wall system. In it, you may, for instance, describe the quality and adequacy of the sheathing which helps make a foundation secure against racking and settling. The siding material may carry a life-cycle guarantee which will also add value to the total reliability picture you develop. Recall from the comparison of resistances that you studied earlier in this chapter that the double-wall system is somewhat better than the single-wall system. In fact, sound and shear factors are among the strongest points in the system. Finally, the reliability has to be high if the best workmanship is to be used to maximum advantage in the construction effort.

In summary, double-wall siding systems usually cost more than single-wall systems. In return, there are distinct advantages, including versatility, adaptability, and minimal bracing requirements.

Siding Varieties

Siding materials are of several types; some can be installed vertically or horizontally; some can only be installed horizontally. Those presently in vogue are panels, boards used with and without battens (S4S, US2S, US4S), lap siding made from various materials and in various ways, and the shingle variety.

Panels: Those for exterior siding may be made from natural wood products or from various manufactured materials, such as cement and plastic. Each has qualities and limitations. In addition to price, climatic conditions, terrain, and design features usually dictate the selection process. Figures 5–8, 5–9 and 5–10 illustrate examples of panel siding.

Figure 5-8 Hardboard Siding, Chanelled Cedar and Stucco Panels Used as Window Accent *(Photo courtesy of Masonite Corporation)*

Figure 5-9 Skip-Troweled Stucco Panel Siding *(Photo courtesy of Masonite Corporation)*

Figure 5-10 Textured 303 Plywood Stained in Earth Tones *(Photo courtesy of American Plywood Association)*

Board Siding is usually installed vertically and may be either board-and-batten or board-on-board. Figure 5–11 illustrates the board-and-batten technique and Figure 5–12 the board-on-board technique.

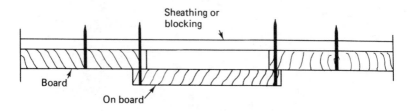

Figure 5–11 Board-on-Board Siding Layout

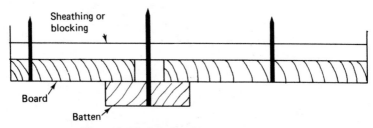

Figure 5–12 Board-and-Batten Siding Layout

Lap Siding has been and is a standard exterior wall covering in many parts of the old and new worlds. Some well-known varieties are S4S (stock 1×8, 1×10 or 1×12), bevel, bungalow (bevel), dolley varden (rabbeted lower back edge), log cabin, and drop. Each is installed horizontally so that the overlaps occur downward in the direction of rain as it falls against the wall. Exterior and interior corners may be self-returned or use corner boards. For self-return corners the siding must be beveled to meet as shown at the left in Figure 5–13. Metal corner pieces slipped over the siding may also be used. See Figure 5–13, right. Figure 5–14 shows a typical lap siding installation.

Shingle Siding is always installed horizontally and lapped downward in the direction of rainfall. The standard $\frac{3}{8}''-\frac{7}{16}''$ cedar shingle, the $1\frac{1}{4}''$ split cedar shingle, asbestos-cement, stained, and even asphalt or vinyl impregnated asphalt shingles are all used as siding. They all require the same installation principles. Namely, that the rows be uniformly spaced and all vertical joints be staggered. Be sure to follow manufacturers installation instructions. You should pay particular attention to the proper type and size of nail to be used. Where strong wind forces and exposure to severe weather are normal, ringed non-

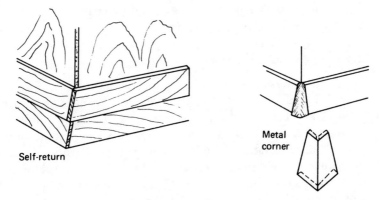

Self-return

Metal corner

Figure 5–13 Corners for Lap Siding

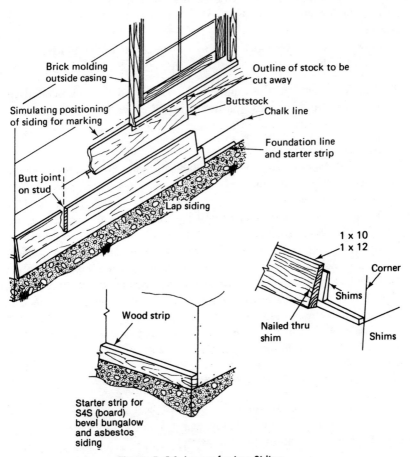

Brick molding outside casing

Simulating positioning of siding for marking

Butt joint on stud

Lap siding

Outline of stock to be cut away

Buttstock

Chalk line

Foundation line and starter strip

1 x 10
1 x 12

Corner

Shims

Nailed thru shim

Shims

Wood strip

Starter strip for S4S (board) bevel bungalow and asbestos siding

Figure 5–14 Layout for Lap Siding

corrosive common nails rather than standard steel nails are recommended.

Estimating Coverage for Siding

Several manufacturers of siding products specify the number of square feet of wall that can be covered with a bundle, a square, or a board foot. Some manufacturers do not, however; so Table 5–1 provides you with a ready reference for the coverage factor of most siding materials. To use this tabulated information, you must work out some data, including: 1) total square feet of wall to be faced; 2) subtraction of square footage or allowance for window and door openings; 3) addition of waste factor or spoilage allowance of say 5% or 10%; and 4) type of siding that will be installed. Computations 1, 2, and 3 give the total square footage of wall to be covered and the table provides the coverage factor which includes overlap, actual dimensions, and general exposure.

Let's work one trial problem to be sure you understand the effectiveness of the table.

PROBLEM: Calculate the number of boards and battens needed to cover the front of a duplex apartment building with a vertical board-and-batten pattern.

COVERAGE DATA:

Wall height 9′8″; length 86′0″. Siding specifications: 1×8 cedar, 1×2 battens of cedar.

PROCEDURE:

1) Round off board length to 10′0″ to establish a standard length with minimal waste.

2) Next, find in Table 5–1 that the batten-on-board coverage factor for 1×8 is 1.0. This means that for each board installed when spaced $\frac{1}{2}$″ to $\frac{3}{4}$″, the actual coverage is 8 inches. Therefore, we multiply the length of the wall by 8″ or $\frac{2}{3}$ feet to obtain the number of boards.

3) Lastly, find in the same table for 1×8 that the number of battens needed is "3 per 2 ft. $+ 1$"
    86′0″ ÷ 2 = 43 units
    43 × 3 battens + 1 = 130 battens

SOLUTION: Estimate for order is 129–1×8 in × 10 ft cedar + 7–1×8 in × 10 ft cedar (5% allowance for waste). 136–1×2 in × 10 ft cedar batten (5% allowance for waste).

TABLE 5–1: ESTIMATING COVERAGE FOR SIDING

Type Siding	Nominal Stock Size	Average Width Overall	Exposed	Coverage Factor
Panels	4 x 8	48"	48"	1.0
	4 x 9			
	4 x 10			
Board-on-Board [a]	1 x 4	3-1/2	2-3/4	1.45
	1 x 6	5-1/2	4-3/4	1.27
	1 x 8	7-1/4	6-1/2	1.2
	1 x 10	9-1/4	8-1/2	1.2
	1 x 12	11-1/4	10-1/2	1.14
Batten-on-Board (Board)	1 x 4	3-1/2	2-1/2	1.0
	1 x 6	5-1/2	4-1/2	1.0
	1 x 8	7-1/4	6-1/4	1.0
	1 x 10	0-1/4	8-1/4	1.0
	1 x 12	11-1/4	10-1/4	1.0
Batten-over-Board [b]	1 x 4	1-1/2	1-1/2	3/1 ft + 1
	1 x 6	1-1/2	1-1/2	2/1 ft + 1
	1 x 8	1-1/2	1-1/2	3/2 ft + 1
	1 x 10	1-1/2	1-1/2	5/4 ft + 1
	1 x 12	1-1/2	1-1/2	1/ ft + 1
Tongue-and-Groove	1 x 4	3-3/8	3-1/8	1.28
	1 x 6	5-3/8	5-1/8	1.17
	1 x 8	7-1/8	6-7/8	1.16
	1 x 10	9-1/8	8-7/8	1.13
	1 x 12	11-1/8	10-7/8	1.10
Shiplap (dropsiding, rustic 3/8" lap)	1 x 6	5-1/2	5-1/8	1.17
	1 x 8	7-1/4	6-7/8	1.16
	1 x 10	9-1/4	8-7/8	1.13
	1 x 12	11-1/4	10-7/8	1.10
Lapsiding (bevel, board bungalow, 1" lap)	1 x 6	5-1/2	4-1/2	1.33
	1 x 8	7-1/4	6-1/4	1.28
	1 x 10	9-1/4	8-1/4	1.21
	1 x 12	11-1/4	10-1/4	1.17
Asbestos @ 27 pcs per bundle	12 x 24	24	10-1/2	2 bundles per square (100 sq ft)
Cedar shingles	3/8" x 18"		Random 8	3 bundles per square
Asphalt-vinyl shingles	12 x 36	12	5	3 bundles per 100 sq ft

a. Based on 3/4" overlap
b. Based on 1 x 2 batten; board widths as listed; 1/2"-9/16" batten overlap

In relation to the above estimating problem: do not forget that if large expanses of doors and/or windows were a part of the wall area, the square foot area of these units should be subtracted from the total square footage of the wall before estimating siding requirements.

Almost every siding job must include some method of finishing the wall-roof joint. A variety of cornices is often used for this area. The closed or boxed cornice style is usually employed for commercial structures.

Cornices

The standard closed or boxed cornice with ventilated soffit, fascia, and trim (molding) will usually resemble residential finishes when used on townhouses and small offices resembling houses. Elements such as lookouts, lookout ledgers, fascias, and soffits are used (Figure 5–15).

Because the length or run of the cornice may be extremely long special attention must be given to maintaining a true line. The quality of the joinery must be excellent and only top quality materials should be used. Siding of a lesser quality may not pose problems, for instance, while poor quality cornice materials and unskilled installation will probably put the job guarantee in jeopardy. There is no way to hide a crooked roof line, or split boards, or sloppy joints, or the soffits that sag because there are no lookout supports.

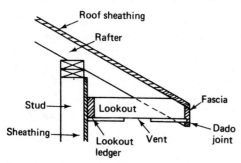

Figure 5–15 Standard Closed Cornice Construction

A different approach to cornice construction is used for commercial buildings. First, the soffit area may range from one to six feet wide. The soffit probably will not have ventilators but will be likely to have lighting. The materials used to make the soffits may include panels, tongue and groove, or metal mesh screening.

Secondly, what would normally be the fascia may be four or more feet high. Figure 5–16 is an example of a vertical 4-ft high fascia in use. The material used as siding for this cornice was manufactured

Figure 5-16 Commercial Cornice *(Photo courtesy of Masonite Corporation)*

by Masonite Corporation. Standard wood, roof or wall framing or trusses provide the necessary support. The usual installation practices recommended by Masonite for joinery and corners were followed by the contractor.

Some fascias are sloped slightly and resemble a very steeply pitched roof. Many national restaurant chains use this technique to maintain a distinctive building style. The fascia in such installations may be made from shingles, lap siding, or board and batten, and from metal and panels in addition.

The installation method, especially of fascia material, is dictated by the siding selected. For some types the single-wall siding technique is appropriate; for other types the double-wall technique must be used. Usually the vertically installed types can be applied with single-wall techniques. Some horizontally installed types can also be installed with the single-wall technique, especially if framing is 16″–24″oc.

With asphalt shingles, however, the double-wall technique must be used since a sheathing must first be installed so the shingles have a proper foundation.

Soffit installations present no serious problems. Extended ceiling joists, truss lower chords or lookouts spaced 16″ or 24″ oc will usually be made a part of the planned construction techniques. The soffit material may require special handling where end joints are inserted into metal channels, for instance. These channels may also be used effectively to compensate for irregularities in the wall or fascia and to reduce the manhours that would normally be required for fitting.

Use of moldings should be held to a minimum ordinarily. They represent costly items and require considerable installation time. Pregrooving the siding or undersides of the fascia can not only be done quickly at the bench but it also eliminates the need for moldings.

Flashings will probably be required on most commercial cornices. This work may be subcontracted if caps, crowns, or special angles are specified. If simple flashing is required the carpenter can incorporate the installation in his general finish work.

Door and Window Unit Installation

Exterior door units used in commercial structures are usually identical to those used in residential construction, especially for the townhouses and small office buildings which resemble houses. They may be installed over either concrete or wood floors and in either single- or double-wall siding systems. Except that there is door sill work, the window installations parallel the door installations.

Sill to Floor Preparation: Over a concrete floor the area is prepared by removing the wooden sill from the door frame unit. After the sill is removed, the beveled-dadoed jambs must be cut square at such a length as to be approximately one-half inch below the bottom of the door. To prevent the jamb members from moving, a spreader must then be cut to length and installed between them and one foot up from the bottom of the jamb.

Preparation of sill to floor area over a wooden floor presumes that a floor joist unit has already been constructed. Ordinarily the sill on the bottom of the door frame is retained, but the subfloor and joist area must be prepared.

Customarily, a cutout similar to that shown in Figure 5–17 is made and, since the subflooring is cut away and some of the joist material is trimmed off, reinforcement is essential. It should be made from 2″

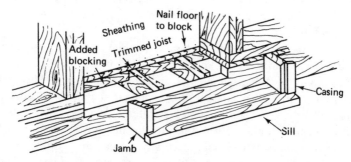

Figure 5–17 Layout for Fitting Door Unit into Wood Joist Unit

stock. Cut, fit, and nail it securely under the edge of the subflooring and under the sill.

Doors in a Single-Wall Siding System: With sill preparations completed, the door unit is ready for installation. Recall that a single-wall siding system uses no sheathing and generally uses a panel siding. This means that the door unit will ordinarily be installed *over* the siding so water proofing must be a primary concern, as well as proper installation.

Pre-fit the unit in order to ensure an adequate nailing surface and correct sill placement. Remove any obstructions that might prevent the inside surface of the door trim from seating against the siding. With this work done, first quality caulking should be installed on the back side of the trim. It is especially important on the door head because flashing is not generally used. So caulking must be tight to keep rain from getting behind the head and causing leaks.

To install the unit, position it and face-nail it through the trim into the studs through the siding. Follow up by applying another bead of caulking around the joint of siding and trim.

Doors in a Double-Wall Siding System: Although the approach described above could be used for a double-wall system, the types of siding usually preclude its application. The more common approach is to install the door unit over the sheathing.

After the sill area preparations are completed, insert the door unit into the opening and check for fit. If adequate nailing surface is available and if it can be made plumb, remove the door and install building paper around the opening. Traditionally the practice has been to fold a 12″ wide strip and nail it so the folded edge is even with the door opening. This provides a quality seating between the frame's trim and sheathing, of course. It also provides a 3″ or 4″ drainage surface for any water that seeps between the siding and trim.

With the door unit installed in the opening, level the sill and plumb the side jambs. Then face-nail through the trim into the studding through the sheathing. Reinforce the hinge and lock areas also by shimming and nailing through the jamb into studs. Follow up by installing flashing over the head of the door.

Wood-Framed Window Unit Installations: In a single-wall siding system, the installation requires pre-fitting to the opening to ensure that the nailing surface is adequate and that any obstacles are removed, for example, the battens on panel-and-batten siding. Install a good quality caulking by making at least one continuous bead around the window on the back side of the trim. Then insert the unit in the opening, raise it to the head, and let it slide $\frac{1}{4}''$ to $\frac{1}{2}''$ down. Then plumb and level and nail through trim and siding into the studs. Follow up with another application of caulking in the siding-trim joint.

In a double-wall installation, a wooden window unit is fitted like a door unit. Building paper is then installed around the opening and, once it is plumb and level, the unit is nailed in place. Finally, the flashing is installed across the top of the window.

Aluminum Window Unit Installation: Aluminum windows are manufactured in all styles. They are single and double hung, casement, awning, hopper, and sliding. Some are shown in Figure 5–18. In each type a flange is formed around the unit to make installation simple. Either of two installation techniques may be employed: the preferred one is under the siding, the other is over the siding. We shall see how each method works with the two siding systems.

To install an aluminum window in a single-wall system the unit must be installed either on the studs or over the siding.

Before installation on the studs, building paper is often laid down. While the paper is not essential for weather proofing it does provide a surface that smoothes out framing irregularities. Whether or not it is used, the window unit is inserted in the opening, raised to the head and lowered $\frac{1}{4}''$ to $\frac{1}{2}''$. Then it is made plumb and level and nailed. Then siding is installed around the window in the regular way.

In the second method, the single-wall siding is presumed to be installed so the window must be installed over the siding. Before the actual installation, the window is tried for fit in order to ensure the adequacy of the nailing surface and to remove all obstructions. A bead of caulking should be installed either on the siding around the opening or on the back side of the window's flange. Then the window is inserted in the opening and nailed.

Since the flange shows, it must then be covered. For example, a brick molding may be cut to fit around the window. Such a molding is cut with 45 degree miters at the head.

Below the sill a piece of stock narrower than the molding must be installed to ensure proper drainage. After that, install the trim and caulk the entire joint area of siding and trim.

Note: Unless the design effect is needed, this second method should be avoided since it is costly in both time and material.

Aluminum window units are easily installed in a double-wall siding system. They are installed over the sheathing and before the siding. Ordinarily you will encounter few if any problems during installation. Insert the window, raise it to the head, lower it slightly, and when plumb and level nail.

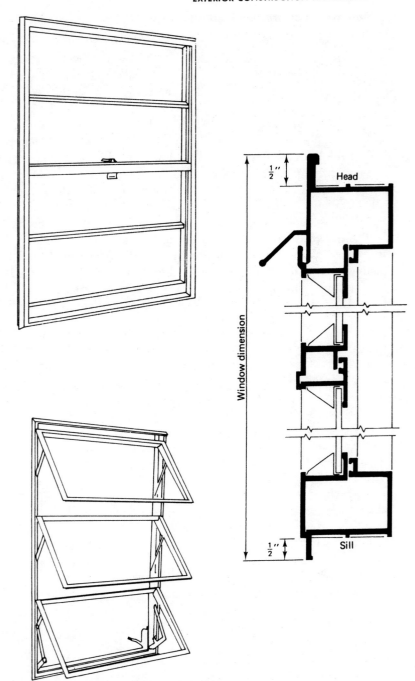

Figure 5–18 Typical Aluminum Window Units; Layout in Profile

Store Front Construction Details

Within the scope of the commercial operations described in this textbook, there are three methods that can be used to construct store fronts. They are standard ways to make the all-wood front, the wood and aluminum front, and the all-aluminum front. The carpenter's role is dictated by the type of front used.

All-Wood and Glass Fronts: Many real estate, insurance and field offices, as well as some retail outlets have the all-wood front incorporated in the building design. The window units are very similar to a large oversized picture window. Therefore, conventional framing is used and, since the units are windows, standard single-wall or double-wall installation techniques are followed. Doors and frames are treated as residential installations are and are not usually a problem.

The all-wood front may be used in a cement wall. If this is the case, nominal 2-inch stock will usually have been placed around the openings during the time the concrete blocks were laid (Figure 5–19). Often enough this stock is used for the window frame since the material is cut, assembled, and installed by carpenters as the wall is

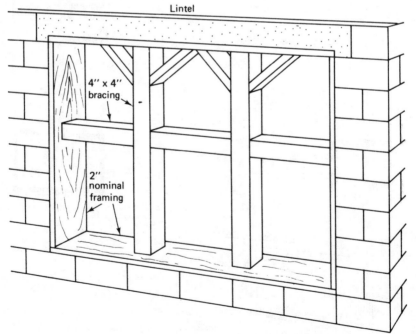

Figure 5–19 Framing Layout for Store Front during Wall Construction

erected. This method provides a base for further window and door development as well as guides for masons. A select grade of nominal one-inch stock is used to build a jamb over the framing. Over this two sets of stops are cut and fitted. One set is nailed securely, the other is tacked temporarily until the glass has been installed.

Many varieties of finish can be given to this type of construction. The members may be finished with stain, varnish, or paint. Siding materials may also be used to cover any exposed framing. Consider the store front shown in Figure 5–20. It is an example of construction that incorporates see-through display window units with an exterior finish utilizing Masonite brand products. Either wood framing or block framing was probably used as the basis for this front and the window units were built of ¼" hardboard panels and applied to complete the job.

Figure 5-20 Store Front Employing Man-Made Masonry and Pre-finished Hardboard Paneling *(Photo courtesy of Masonite Corporation)*

Combination Wood and Aluminum and Glass Fronts: The only major difference between the wood fronts just discussed and the combination front is the inclusion of aluminum window units. Figure 5-21, a closeup of a restaurant, provides an excellent example. Conventional framing was used along with Masonite brand textured panel siding and rough-sawn cedar siding. A glass company installed the aluminum window units.

Where subcontractors, such as glaziers are used, close coordina-

Figure 5-21 Combination of Materials and Aluminum Window Frames *(Photo courtesy of Masonite Corporation)*

tion is necessary. Since the window frames are often installed before exterior siding and trim are applied, it is particularly important that the carpenters prepare accurate openings—openings that are square, plumb, and level. The carpenters are also responsible for the accurate fit of all the siding and trim to the metal frames installed by the glaziers.

Combination fronts can also include the aluminum door and frame. The carpenter's tasks are the same as those described for windows.

All-Aluminum and Glass Fronts: These installations will be subcontracted ordinarily. So it is essential that the openings be accurately sized as well as square, plumb, and level. A mastery of such installation carpentry tasks is not limited to specialists. It is important that every workman understand the fundamental techniques of installation. This may well help him to avoid costly errors and will certainly provide a basis for doing a quality installation.

Installation Criteria: A minimal 4″ wall depth is required for installation of the channels that in turn hold the glass. It is also true for the door jamb. The channels are fairly rigid, but they are not strong enough to be used as supporting members. For that reason, headers must usually be installed. But in some installations the ceiling joists extend to the fascia and provide a covered walk area. In such cases the joists themselves are the anchor for the channels. Either support method described above is proper and adequate for head installation of the channel.

Installation Sequence: The routine for installation of metal framing materials includes layout, materials procurement, cutting, and fitting, for both the window areas and door openings. The glazier uses various special fasteners to secure the aluminum frames to the walls, base, and overhead. Following this the glass is installed in the window areas and the door is hung in the jamb. Then accessories, such as locks, handles and clamps are installed.

To Summarize: Exterior completions require the carpenter to have a very wide range of skills and information. He must be extremely knowledgeable about techniques, such as those for single- or double-wall systems and about the structural quality of the products he installs. He must also prepare work to the requirements of others. As we have seen, he must be skilled at forming the openings for doors and windows and store fronts for the masons and glaziers who perform special work. So the carpenter does both preparatory and finish tasks. Where finish work is needed, he must perform with a high degree of accuracy, using all his skill and knowledge.

TRAINING ON EXTERIORS

Since completion of the exterior of a building represents the final phase of outside work, all exterior carpentry skills must be classified as finish, except for sheathing. So the training plan must identify those persons who have had previous training and experience in single-wall and double-wall siding systems. The plan should also include each subphase of work that requires a skill.

Recording the Training

A typical training plan which supports the activities of exterior wall completion is shown in Figure 5–22. Your plan would, of course, only list those skills needed for your job. Accordingly, some of the tasks shown would not be applicable. It may be, however, that you need

Figure 5–22 Training Record for Exterior Construction Crew

to define to a greater degree those skills that are needed. If so, break down each task to the individual skill requirements.

As stated previously, each new carpenter or apprentice who is hired must be interviewed in order to determine his skill in the different subphases of exterior work. In such an interview full responses must be elicited by meaningful questions. For example, "Have you ever installed panel siding?" is a question that does not ask much. The answer will not reveal much either. You won't know if John Doe has installed the siding vertically, horizontally, or in single- or double-wall systems, or whether he glued or nailed or battened or what types of siding he has installed. To develop meaning for your training plan, something like "Explain in brief terms the procedure you would use to install panel siding with a single-wall siding system." will yield useful information. The response will give you some idea of the depth of his knowledge (or lack), level of experience (or inexperience), and understanding (or ignorance) of related tasks that precede, accompany, or follow such an installation.

The same sort of questions about double-wall siding systems, various types of siding, cornice construction, store fronts, and flashing will usually result in answers that define the applicant's overall knowledge and degree of general competence. It will also provide you with a solid basis for planning any needed training program on exterior finishes.

Task Activities

To support your training plan, appropriate task activities have been developed and follow in the next section. They should be used in any training that concerns the exterior siding phase of the total construction project. Each task activity instructs in a segment of the whole phase. Select only those that apply to the present job. Remember, if specific materials can be referred to, each task activity will have more immediate association for those who are either learning or reviewing work.

Remember also that the procedures section can be developed as a check list for training. Similarly, the manhour assignments can be used to gage the time alloted for a given job. Time and procedures are not all that necessary, however. Quality of workmanship is very important.

EX 1 TASK ACTIVITY: SHEATHING WALLS

RESOURCES

Estimated Manhours for 2 men: 10 mins per 32 sq ft (avg.)

Materials:

elevation plans
plywood or fiberboard
nails

Tools:

#8 crosscut saw
chalk line
16-oz hammer
scaffolding as required
6-ft folding ruler
level
portable power saw

PROCEDURES

1. Verify and adjust plumb of walls and then reinstall any needed bracing as a preliminary to sheathing installation. Along with this, install both interior and exterior corners if not previously done.
2. Make plans for either a vertical or a horizontal sheathing installation. Measure and pre-cut panels of sheathing so that ends or edges break on center of studs. Position first panel on the wall by starting from the corner and foundation sill line; nail to studding. Measure, mark, and cut, as required; install the next panel, allowing proper spacing between joints for expansion.
3. Make cutouts for window and door openings either on the ground before installing panel or after it is in place.
4. Complete the wall; then proceed to the second, third, and fourth walls.

EX 2 TASK ACTIVITY: SIDING FOR SINGLE-WALL SYSTEM

RESOURCES

Estimated Manhours for 2 men: 15 mins per 32 sq ft (avg.)

Materials:

elevation and detail plans
siding panels and battens if used
nails

Tools:

#8 crosscut saw	framing square
chalk line	6-ft straightedge
6-ft folding ruler	level
portable power saw	portable power drill
set of spade bits	portable power vibrating saw
scaffolding as required	

PROCEDURES

1. Prior to installing paneling, verify accuracy of the walls for plumb; make required adjustments. Complete any interior and exterior corners that were overlooked.
2. Lay out siding for installation on the wall by establishing that position of first stud is not farther than 48″ from corner. Verify that there is another stud at 48″ span. Finally, verify that last piece to be installed is more than 6″ wide. If not, go back and lay out first piece for a less than 48″ span and repeat above steps.
3. Measure, mark, and cut first piece for width and length. Length, as a rule, should be from ½″ below sill/foundation joint to slightly above the lookout ledger or even with the roof's edge. Measure accurately, then lay out and cut out window and door openings as needed.
4. Position prepared panel. If it is ripped to width on site, the cut edge should be on the corner and factory edge on the center of the stud. Nail in place according to nailing schedule.

5. Repeat the above procedure for at least one more panel; then if battens are to be used, install them after each panel installation. This eliminates excessive movement of scaffolding.

6. Repeat the above procedure for all walls. (Butt joints are needed for walls whose height exceeds the panel height. Refer to pages 138–139 and Figure 5–6.)

EX 3 TASK ACTIVITY: SIDING FOR DOUBLE-WALL SYSTEMS

RESOURCES

Estimated Manhours for 2 men: 1 hr per 60 sq ft (avg.)

Materials:

elevation and detail plans
siding
building paper if used
starter strips if used
nails
caulking
flashing

Tools:

#8 crosscut saw	framing square
14-oz hammer	bevel square
6-ft folding ruler	set of wood chisels
combination square	tin snips
level	chalk line

PROCEDURES

A. *VERTICAL INSTALLATION*

1. Lay out the wall for the siding. If panels are to be used, locate on-center studs and intermediate studs. Again, ensure that last piece on wall is more than 6″ wide. If less, start first piece one stud back. If board-on-board or board-and-batten siding is to be used, install horizontal blocking at 24″ oc between studs. Chalk a line on the outside of the sheathing where the blocking is located.

2. Measure, mark, and cut the first panel or piece of stock. Nail it in place. If it fits well, pre-cut several pieces at one time.

3. Since window and door units will be installed prior to the siding, mark and cut siding to fit accurately around these units. Do not force pieces in place; instead allow 1⁄16" to 1⁄8" space for expansion. Where required, install flashing before applying siding above doors and windows.

B. *HORIZONTAL INSTALLATION*

1. Install building paper if to be used; then cut and install the starter strip material. The starter strip's lower edge should be flush with the sill/foundation joint.

2. Lay out a pattern on the wall to determine spacing and overlap requirements for siding courses. Wherever possible, adjust spacing so the bottom edge of a course breaks even with each window sill and the top edge of a course extends 3⁄4 " above each window rain cap.

3. If corner boards or metal corners are to be used, install them.

4. Measure, mark, and cut the siding stock to fit the corner and to break on-center of a stud. Nail it in place. Either complete the course or begin the next several courses, being sure to break joint each time.

5. When fitting around window and door frames, do not wedge pieces in place; instead cut them so as to allow for expansion. Complete the installation by fitting the frieze board over the last course of siding and under the lookouts or roof edge.

EX 4 TASK ACTIVITY: CORNICE ENCLOSURE STANDARD AND COMMERCIAL

RESOURCES

Estimated Manhours for 2 men: 1 hr for standard 12 lin ft
1 hr for commercial 4–6 lin ft

Materials:

elevation and detail plans
cornice stock (fascia, soffit, frieze, trim)
nails and securing devices
sheathing and/or furring

Tools:

#8 crosscut saw
chalk line
level
ladders
14-oz hammer
6-ft folding ruler
scaffolding
portable power saw

PROCEDURES

A. STANDARD

1. Lay out cornice area so as to establish both the position of the lookout ledger against the wall and the final height of the bottom lower edge of the rafters. Trim back to the reference line any rafter lower than the reference line. Install a lookout ledger and lookouts.

2. Prepare fascia boards to proper width and make groove for the soffit on the back lower edge. Install the fascia.

3. Prepare the frieze board to overlap the last course of siding (even up brick siding) by ½" and install.

4. Trim soffit material (commercially prepared for ventilators) to proper width if plywood is used. Where perforated metal soffit is used, install per manufacturer's instructions.

5. Install either a 1 × 2 or ¾" quarter-round molding in the joint between frieze and soffit.

B. COMMERCIAL (FRAMING COMPLETED)

1. If required, apply sheathing or furring to entire height of fascia area with material specified. Establish a lower edge overhang either by using data in detail drawings or by calculating the requirement.

2. Either pre-cut fascia stock to correct lengths for vertical installation or pre-cut stock length to break on-center of framing. Install, using method that minimizes nail exposure.

3. Assuming that lighting is already installed in soffit area, install soffits. If a frieze board is planned, cut to fit and install against the wall and up to the soffit.

EX 5 TASK ACTIVITY: SETTING EXTERIOR WOODEN DOOR UNITS

RESOURCES

Estimated Manhours for 1 man: 1 hr per jamb

Materials:

elevation and detail plans
door jamb
blocking
15-lb or 30-lb felt or caulking
flashing
casing nails, common nails
metal threshold if required

Tools:

#8 crosscut saw
combination square
set of wood chisels
6-ft folding ruler
crow bar (small)
power saw (*optional*)
electric drill and bits (*optional*)

14-oz hammer
framing square
level
brace and set of augers

PROCEDURES

A. *ON A WOOD FLOOR*

1. From plans obtain finished sill height; then lay out work area. Determine area of subflooring to be removed, lay out, remove nails and cut away subflooring.

2. Mark joists and box beam to a slope and depth that will provide secure, firm foundation for door sill. Trim excess stock and try jamb for fit; make adjustments as required.

3. Before installing jamb, fit weatherproofing felt paper or caulking around the opening. Install the jamb shim sides, being sure to maintain plumb and keep square; face-nail through brick molding or exterior trim into studs.

4. Fit and hang the door.

B. ON A CEMENT FLOOR

1. Remove sill and prepare the jambs for insertion into door opening. Trim beveled ends of side jambs so that when cut square they are ½″ to ¾″ longer than the door height.

2. Install a spreader 12″ up from the trimmed ends and long enough to maintain the uniform separation of the jamb sides from top to bottom.

3. Position on the head jamb in the opening and try for fit. Remove and prepare surfaces with weather material; insert jamb and when plumb and level, face-nail through the trim.

4. Insert shims between side jambs and studs or the blocks of wood set in the cement and nail through the jamb with casing nails.

5. Cut and install drip cap flashing as required.

6. Fit and hang the door; install metal threshold under it.
   NOTE: When this type of unit is used in a single-wall siding system, it is installed over the siding and usually before any battens are nailed in place. In a double-siding system the unit is usually installed over the sheathing instead.

EX 6 TASK ACTIVITY: WINDOW UNIT INSTALLATION

RESOURCES

Estimated Manhours for 2 men: 10–20 mins per unit (avg.)

Materials:

elevation plans
window schedule
caulking or felt paper (as required)
nails or self-threading screws

Tools:

level   ladder or work bench
14-oz hammer

PROCEDURES

1. Select the unit (wood or aluminum) for the wall opening according to the window schedule. Insert unit into opening from outside; verify accuracy of the opening and adequacy of the nailing surfaces, especially at head and sides.

2. Prepare the unit for installation: for single-wall siding systems caulk the back side of the flange or trim; for double-wall siding systems install building paper around the opening on the sheathing.

3. Then carefully insert the unit into the opening as close to the head as possible; lower it from the head ¼" to ½" and when plumb and level, nail.

4. If a molding/trim is to be installed over the flange, prepare the pieces, using 45 degree miters on corners at the head; and nail with 8d casing nails.

EX 7 TASK ACTIVITY: PREPARING STORE FRONT FOR GLAZING

RESOURCES

Estimated Manhours for 2 men: 4–16 hrs per opening

Materials:

detail and elevation plans
2" nominal stock
1" nominal stock, clear and/or select grade
nails and other fasteners
trim and molding

Tools:

#8 and #10 crosscut saws
14-oz hammer
level
bevel square
sawhorses

combination square
portable power saw
6-ft folding ruler
step ladder

PROCEDURES

1. Using detail plan, determine extent of the finished glassed area; prepare openings accordingly, using 2″ nominal stock where framing is required.

2. Cut and install the 1″ nominal stock to form jamb and sill area; make all cuts using the butt joint.

3. If aluminum and glass units are to be used, complete the interior after they are installed.

4. If glass is to be installed in a wood frame, prepare it for glazing. Note that this is usually the 1 × 2 or 1 × 3 interior stock with the glass installed outside.

5. Cut and install outside trim around the opening; use caulking and flashing as required to weatherproof the entire area.

REVIEW QUESTIONS

1. What are four of the major concerns in the planning phase of exterior completions?

2. What makes up a single-wall siding system?

3. What are several cost factors to consider when using the single-wall siding system?

4. Is the reliability of a single-wall siding system acceptable by FHA standards?

5. What is the composition of a double-wall siding system?

6. How can a close study of specifications be translated into cost savings and certification of a guarantee?

7. How should caulking be used when siding is installed?

8. Describe the difference between board-on-board and board-and-batten siding systems.

9. What are the parts that usually make up a closed cornice?

10. If asphalt strip shingles or the like are used on a commercial cornice, would a single-wall or a double-wall siding be more appropriate? Explain.

11. Why is the under-door sill and floor support so essential to an exterior door frame installation?

12. What are the different procedures for installing an exterior door unit in a single-wall and double-wall siding system?

13. Describe the work the carpenter must do when he constructs the all wood and glass store front.

6

Interior Tasks and Techniques

Acoustical: dealing with and pertaining to the science of sound; (of a building material) designed for controlling sound in a structure
Cellular trim: vinyl trim for doors, baseboards, and edges of panel
Deadening board: a manufactured panel whose sound absorbing properties lessen or suppress noise
Fiber blankets: either glass fiber or mineral wool building blankets used for thermal control in walls, floors, and ceilings
Fire control: the combination of methods and materials used in construction to retard and contain fires
Gypsum wallboard: a wall panel made from gypsum (soft plaster-like mineral) in a binding agent so as to be firm enough to make sheets covered with paper; fire-retarding wallboard also has vermiculite mixed

with the gypsum and is covered with chemically treated papers
Non-load bearing: said of an intermediate wall or any member that does not carry a load
Resilient channel: a metal channel or three-sided rectangle used in horizontal lengths over framing and to which kinds of sound-absorbing and fire-control materials are attached
Soundproofing: a method of applying a composite of insulating and sound-materials within and to walls, floors, and ceilings so as to reduce noise transmission to prescribed levels
Vermiculite: the particles of one or other of the minerals used in fire-coded gypsum panels because of their ability to expand greatly in the presence of heat and thus trap and retard fire

With the building closed-in, a variety of interior work activities can be scheduled for completion. They include both highly skilled tasks and general tasks. Furring of walls, installation of gypsum panels, and other general tasks can be performed by carpenters with a minimum of experience. Short training sessions will usually result in reliable and significant productivity.

As with exterior carpentry there are many activities, however, that require the hand of a master craftsman. Installing interior door units, trimming rooms, installing drop ceilings, and paneling walls are a few. Trimming is the most exacting work of all.

Data are provided in this chapter on the above products and on their installation techniques. Even though installation of gypsum panels, drop ceilings, and flooring may sometimes be subcontracted, it is important to understand their characteristics and the installation techniques. Additionally, tasks activities are developed for these products in the training section.

Many of the new products used in commercial buildings are pre-finished, so the craftsman must use extreme care during fitting and installation to prevent marring the surfaces. Because of this, several specialized installation techniques have been developed. But first, let's view the organizational function in this phase of the job.

ORGANIZATIONAL FUNCTIONS IN
INTERIOR CONSTRUCTION

Except for the initial phase, planning effective integration of the work on building interiors is complicated because many subcontractors are involved. Figure 6–1 shows that a contractor might subcontract almost every type of work that needs to be performed. If he chooses to do this, his crew will likely be small and very specialized. They will be classed as *finish carpenters* and will install door units, trim, and pre-finished paneling, and do the ceiling work.

The data developed by the scheduler is used by the contractor during the conference and negotiation period with subcontractors. Final dates and duration of work activities can be scheduled and any foreseeable delays can be identified. Then alternate plans can be developed.

Of course, any alternate plan must be carefully developed. It is a compromise; therefore, there may be extra cost factors arising from delays, substitution of different materials, and requirements for additional crewmen.

There is a parallel between the sequence of work performed on commerical buildings and residential buildings. Work sequences follow accepted procedures. Tasks involving the walls, floors, and ceilings are done first. Walls are insulated and fireproof materials are added. Ceilings are installed and insulated, and the finish trim work is done next. Floors are usually installed last except for the T&G hardwood floors.

The production foreman must participate in the planning phase of

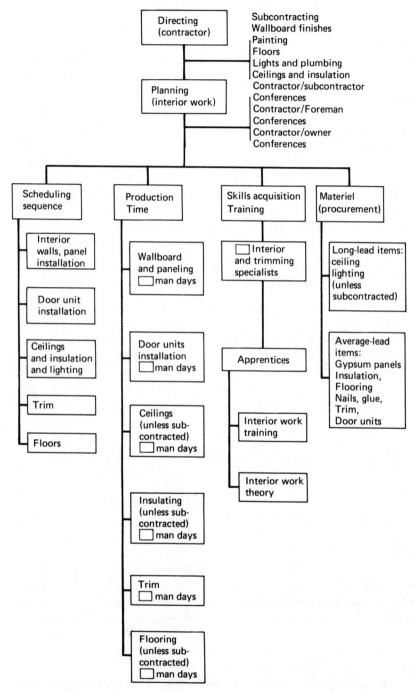

Figure 6-1: Organization of Interior Construction Activities

the interior just as he has in all other phases. Once he is aware of the scope of work expected of his crew, he can determine the adequacy of their skills and any training requirements. He will also make plans to secure additional workmen if required. He can also make preparations for subcontractors so they will not encounter delays which may prove costly to everyone concerned.

An example of careful planning can be seen in outline in the following case: *Store XYZ* was completed on time. The contractor had planned so that his crew and the electrical and heating subcontractors were all able to start at the same time. Paneling and fire resistant materials were prepared for installation while outlets, lighting wiring, and ducts were installed. Two days later the insulation subcontractor and wallboard subcontractor began work as scheduled. After that, the ceiling was blow painted by the painting contractor as planned. Then he went on to finish the walls. (If a drop ceiling had been specified, the wall panel would have been installed first.) With walls and ceiling completed the carpenters installed doors, and trim. Next, still according to plan they completed the display window platforms. Then, after all the painting was done the floors were installed. *Store XYZ* was ready for fixtures and goods.

Involved in the example above are various decisions that must be included in the planning phase. Types of materials, types of structure, and methods of work are all important elements that make each plan unique to the particular job. While there is a general, logical sequence, each job also has a specific sequence that can only be developed by careful planning.

It is necessary to study interior construction techniques so that proper decisions can be made during the planning phase and so that details of various tasks can be well defined for performance.

THE INTERIOR CONSTRUCTION

A detailed understanding of the tasks and the variety of application can start from what we know. Townhouses and small office buildings are classified as commercial buildings for construction, but their interiors are completed by using many of the same construction techniques that are used for residences. Let's begin with wall completions.

Interior Walls

With the construction framing phase already done, the basis for the walls is prepared. During that phase provisions were made for adequate corner supports and for the later construction of sound-deadening and fire-retarding partitions. Now work starts with the application of the sound-deadening insulation between the walls and standard insulation on exterior walls for heating and cooling radiation control. Recall that an acceptable sound transmission rating is near 50 and that higher ratings of 55 to 60 or near optimum can be achieved.

Completion of wall work should be thought of in terms of wall systems—systems with characteristics or elements, such as building materials to be used, installation methods, sound proofing requirements, and fire control minimum standards. Four such systems are commonly used in commercial applications. They are examined next.

SYSTEM 1: 2-hour rated wood-stud partition, STC 50 or better[1]

A. Standard: The wall or walls must be closed with materials that can resist fire for two hours and have a sound transmission quality of STC 50 or better. This system uses standard wall framing with studs and plates.

B. Possible solutions:

1. 2 layers 5/8" fire rated gypsum drywall on each side; 2 X 4 16" o.c. — 3" sound attenuating, fire rated blankets — RC-1 channel one side 24" o.c. — resilient side screw attached, opposite side nail attached — both base layers applied vertically and face layers applied horizontally — resilient layers perimeter caulked, joints finished. STC 58.

2. 2 layers 5/8" fire rated gypsum drywall — 2 rows 2 X 4 16" o.c. on separate plates 1" apart — base layer attached with 6d coated nails 16" o.c. — face layer attached with 7d coated nails 7" o.c. — perimeter caulked — joints finished. STC 51; With 3½" thick blanket in one cavity STC 56. (2 hours estimate).

Each solution provides the method of installation within its description. The builder need only understand the materials indicated and he or she will understand the construction techniques required.

[1]USG Drywall/Wood frame systems, SA 924, United States Gypsum Company, Chicago, IL; 1987, page 2

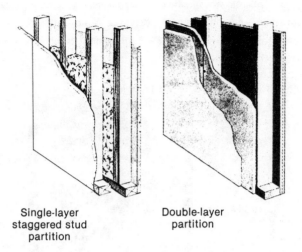

Single-layer
staggered stud
partition

Double-layer
partition

Figure 6-2: Wood Drywall Systems, **Courtesy of U. S. Gypsum Company**

SYSTEM 2: 1-hour rated wood-stud partition, STC 50 or better.

A. Standard: Materials for this wall closing must have the ability to resist fire for one hour and have a sound transmission control quality of STC 50 or better.

B. Possible solutions:

1. Resilient partition - 5/8″ fire rated gypsum drywall — 2 × 4 16″ o.c. — 3″ sound attenuating, fire rated blankets — RC-1 channel one side spaced 24″ o.c. — panels attached with 1″ Type S screws — opposite side direct attached with 1¼″ Type W screws — joints finished, perimeter caulked. STC 50.

2. 5/8″ fire rated gypsum drywall — 2 × 3 non-load bearing studs 16″ o.c. — 2 × 3 plates 1″ apart — panels nailed 7″ o.c. — 3″ sound attenuation, fire rated blankets, one side — joints finished — perimeter caulked. STC 54.

3. 2 layers ½″ fire rated gypsum drywall each side — 2 × 4 16″ o.c. — 3″ sound attenuating, fire rated blankets — RC-1 channel one side spaced 24″ o.c. — resilient side screw attached — opposite side nailed — both base layers applied vertically and face layers applied horizontally — base layers perimeter caulked — joints finished. STC 58; without blankets STC 52.

Again there is no need to describe the construction techniques since the solutions so aptly describe them. However we supply Figure 6-2, a visual representation of several solutions.

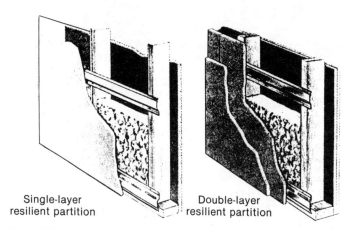

Single-layer
resilient partition

Double-layer
resilient partition

Figure 6-3: Wood Drywall Systems with Resilient Channel, RC-1, **Courtesy of U. S. Gypsum Company**

SYSTEM 3: 3-hour rated USG SJ-stud/RC-1 channel partition, STC 62[2].

A. Standard: Materials for this wall closing must have the ability to resist fire for three hours and have a sound transmission control quality of STC 62.

B. Possible solution: Resilient stud wall — 5/8″ fire rated gypsum drywall — 35SJ20 studs 24″ o.c. — 3″ sound attenuating, fire rated blankets — RC-1 channel one side spaced 24″ o.c. screw to studs — 3 layers gypsum panels screw attach to studs, 2 layers screw attach to RC-1 channels — panels applied vertically with joints staggered — joints finished — perimeter caulked STC 62.

SYSTEM 4: 2-hour rated steel framed drywall, STC 50 or better[3].

A. Standard: Materials for this wall closing must have the ability to resist fire for two hours and have a sound transmission control quality of STC 50 or better.

B. Possible solutions:

1. 2 layers ½″ gypsum fire rated drywall on each side — 3 5/8″ steel studs 24″ o.c. — panels applied vertically and staggered — base

[2]USG steel stud/resilient drywall systems, SA 921, United States Gypsum Company, Chicago, IL; 1987, page 2
[3]USG steel-framed drywall systems, SA 923, United States Gypsum Company, Chicago, IL; 1987, page 2

layer screw attached — face layer strip laminated or screw attached — joints finished — perimeter caulked. STC 50. Include 1½" sound attenuation, fire rated blankets STC 55; Use 2½" studs and blankets STC 54.

2. 2 layers 5/8" fire rated gypsum drywall applied each side — 3-5/8" steel studs 24" o.c. — base layer screw attached, face layer laminated or screw attached — joints staggered and finished or unfinished — 3" sound attenuation, fire rated blankets — perimeter caulked. STC 56.

3. 2 layers 5/8" fire rated gypsum drywall each side — 2½" steel studs 24" o.c. — panels applied horizontal and joints staggered — base and face layers screw attached — 2½" sound attenuating, fire rated blankets — perimeter caulked. STC 51.

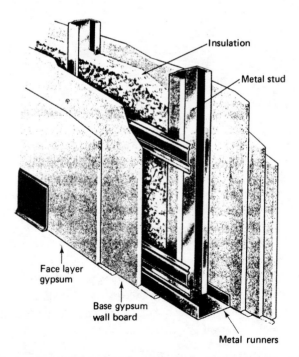

Figure 6-4: Steel Framed Drywall System with RC-1 Resilient Channels, **Courtesy of U. S. Gypsum Company**

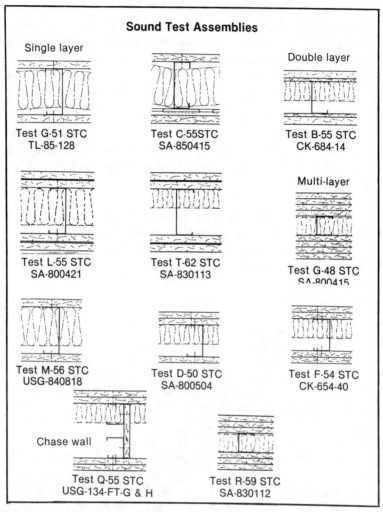

Sound Test Assemblies

Single layer

Test G-51 STC
TL-85-128

Test C-55STC
SA-850415

Double layer

Test B-55 STC
CK-684-14

Test L-55 STC
SA-800421

Test T-62 STC
SA-830113

Multi-layer

Test G-48 STC
SA-800415

Test M-56 STC
USG-840818

Test D-50 STC
SA-800504

Test F-54 STC
CK-654-40

Chase wall

Test Q-55 STC
USG-134-FT-G & H

Test R-59 STC
SA-830112

Figure 6-5: Top View, Steel Wall Systems, **Courtesy of U. S. Gypsum Company**

Steel stud wall systems can be constructed with fire ratings of one hour to 4 hours. Depending upon the number of layers of gypsum drywall, their thickness, and the blankets used for sound and fire control STC values can range from a low of 42 to a high of 62.

Figure 6-4 shows a steel wall section and Figure 6-5 shows several different steel wall systems from the top view.

Conclusions on Fire and Sound Control

Materials and Their Uses: Various fire and sound control systems are capable of satisfying standards by a combination of materials and methods of installation. Fire rated gypsum wallboard is the primary fire retarding element. Mineral fiber insulation blankets, if used, may also aid in fire control although their primary function is thermal protection and sound control.

The materials which make gypsum panels extra fire-resistant are the inclusion of vermiculite and chopped fiberglass. The vermiculite (granules of various micaeous minerals) imbedded in the gypsum expands greatly when heated. This expansion increases the material's volume and tends to block the passage of the fire.

We have seen from the materials specifications that each wall covering system requires the use of fire rated gypsum drywall. These panels may either be the unfinished variety which requires taping and painting or one of the factory-finished vinyl coated varieties which are installed with batten strip joints over the joints and metal corners.

Furthermore, common sense tells us that such gypsum panels must also be installed as base material if a pre-finished wood veneer paneling is to be installed on any wall requiring fire and sound control. The flammable veneer panels are glued to or nailed over the gypsum panel. In a fire the wood paneling would burn, of course, but the fire rated gypsum would contain the fire throughout the rated time period.

Note that various systems calling for sound transmission control of *STC 50,* or better, all meet industry standards with combinations of materials which deaden, absorb, and reduce excessive noise. The combinations may use one or the other framing styles, insulation blankets, more than one layer of gypsum drywall and resilient channels on one side of the wall. With any of these methods an acoustical sealant is used around all wall perimeters and utility outlets to complete the work.

Cost Analysis: No estimating data are given in this text since pricing varies greatly in different parts of the country and particularly reflects demand/supply fluctuations in the construction industry. You must make a cost analysis sometime during the estimating phase of contract negotiations, however; so you will want to examine the job specifications closely in order to define the required standards. They may, for instance, require a single configuration such as a 1-hour rated wall with an *STC* median of *50.* If this is the case, the cost analysis should then include a study, with prices for various systems. Items vital to the analysis include:

1. Quantity and variety of materials
2. Manpower (or man-days) required
3. Dollar value for items *a* and *b*

Such a cost analysis may reveal cost differences for different completion methods that could range from several hundred to several thousand dollars.

If the overall building specifications are variously subdivided so that different walls have different qualifications, a cost analysis must be made for each type of wall system specified. For example, some walls may not require a 1-hour rating for fire control; and some walls may require a 2-hour rating. Then, too, some walls in the same building may not require any sound transmission control.

In summary, cost analysis is required to select the most economical method of completing each wall. It must list the precise type and quantity of materials needed either for the entire job or for designated walls as per structural specifications. It must also break out the manhour costs by wall systems.

Installation of Prefinished Paneling

Learning which wall system to use also opens the possibility to use prefinished gypsum panels. Because of the prefinish surface, different techniques must be used for installation. Also new materials are employed for trim functions. We shall discuss some of these things.

Prefinished panels require the needs of a variety of moldings to make a finished job. Figure 6-6 shows the variety offered by U.S. Gypsum Company. As you can see, each is designed for a special need.

Generally, moldings should be stored at room temperature for 24 hours before installation. Start installation at a corner or door that is plumb and level. Cut moldings with a fine-tooth hacksaw, mitering the same way as with wood moldings. Cut $\frac{1}{16}$" short for a loose fit to allow for thermal expansion; never force moldings into place. Fasten moldings with flat-head wire nails, staples or drywall trim screws 8 to 12 inches o.c. (for snap-on moldings, with drywall nails or bugle head screws through holes in the retainer).[4]

Gypsum panels applied to either wood or steel studs should be installed using the following guidance:

a. Apply 8"-long strip of vinyl foam tape to the facing of each stud, positioned at midpoint of studs up to 8' long; at third points on studs up to 12' long and quarter-points on studs over 12' long. Where no mechanical fasteners are to be used at top or bottom of stud, apply an 8"-long strip of tape [at these points]. Apply a continuous ⅜" bead of drywall stud adhesive to the entire face of studs between vinyl foam tape. Immediately apply [prefinished] gypsum panels vertically and apply sufficient pressure to insure complete contact.

b. [Prefinished] gypsum panels applied over base layer of gypsum panels. Apply liquid contact adhesive to back of [prefinished] panel and face of base layer according to adhesive manufacturer's directions. Allow adhesive to air-dry, then bring panels into contact. Impact entire surface to assure complete contact.

c. [Prefinished] gypsum panels applied over base layer of masonry, gypsum board, wood or mineral fiber board. For interior masonry walls and gypsum board, apply continuous strips of vinyl foam tape to entire width of [prefinished] panel back at midpoint and ⅜" from each end. Spread laminated adhesive over the entire area of panels between tape using notched metal spreader with ¼" x ¼" notches spaced 2" o.c. Position panel and immediately apply sufficient pressure to assure complete contact over entire surface. (Mechanical fasteners at top and bottom of panel.)[5]

Opportunities for Savings: The contractor can realize a cost saving if he is able to complete wall work by using pre-finished paneling. Granted, such work is classed as finish work and, therefore, requires highly skilled workmen. But even the cost of this labor plus the extra cost differential for pre-finished materials is less than the total labor and materials cost for unfinished paneling that requires painting or

Installation Details

| RP-2, RPV-2 | RP-4, RPV-4 | RP-5, RPV-5 |
| Inside Corner | End Cap | Snap-On Corner |

Figure 6-6: Trim for Gypsum Paneling, **Courtesy of U. S. Gypsum Company**

---

[4]Textone vinylfaced gypsum panels, SA 928, U.S. Gypsum Company, Chicago, IL; 1987, page 5
[5]Ibid, pages 7 and 8

Installation Details (continued)

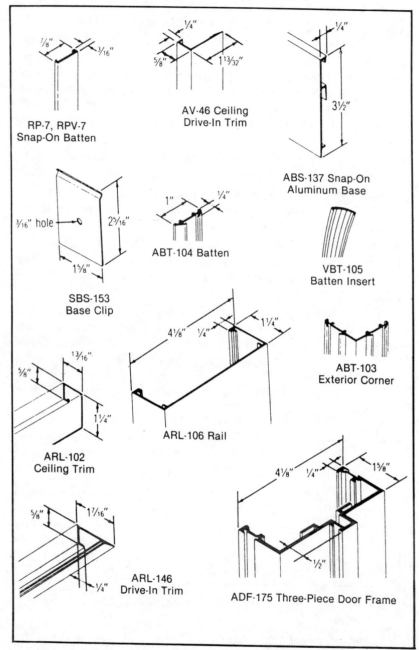

Figure 6-6 (cont.): Trim for Gypsum Paneling, **Courtesy of U. S. Gypsum Company**

paperhanging as well. In addition, completion time is reduced considerably. That in turn gives the contractor an opportunity to obtain a contract bonus if there is one in the contract.

Pre-finished panels also provide a building owner with wide selection of wall finishes. The type, grade, quality, color, and texture of the material can all be decided upon during the contract stage of planning. This gives the contractor considerable lead time in which to place his orders.

Since interior decoration requires the design coordination of walls, ceilings, floors, and furnishings those sub-contractors will also have sufficient lead time for placing material orders.

The use of pre-finished wall coverings also impacts the scheduling and planning of other subcontractors. Those who execute related tasks, such as the lighting or heating installation need to do certain stages of the work before, during, and after the walls are enclosed. Closely interfaced scheduling of these subphases will produce profits for all who perform them. The subcontractor who finishes the walls, ceilings, or the like must integrate his work in the total job. If he paints, for instance, he may do some before the paneling is installed and the finish work after the panels are installed.

Plans that offer chances for savings are closely related to management's total concern for the job. If the work is to be well done and if profits are to be made, a well planned, coordinated effort is essential. It is also imperative that the contractor's employees and subcontractors cooperate with him and with one another in this phase, as in every phase of commercial carpentry.

Ceiling Coverings

Ceilings are the overhead interior linings of rooms and in commercial buildings are usually installed with materials and methods similar to those used for residences. There are a few differences, however, and we shall discuss them. Generally, a commercial ceiling is finished with acoustical materials by a method called "dropped" or suspended or is *blown* with a paint and joint compound mixture over gypsum panels or surfaced with wood. Occasionally, though, a level flat ceiling may be taped and wallpapered or even glued to a base underlayment made from fire retarding gypsum paneling. In order to learn the technical characteristics of each of the three general types, let's examine each as a system.

SYSTEM 1: Suspended insulated ceiling of acoustical panels with class A fire rating as specified by FED SPEC SS-A-1186 and with sound control mandatory.

    A. STANDARD: This system requires a thermal, sound deadening, and fire retarding installation. Insulation with glass fiber blankets and the rated ceiling panels will meet both sound control and fire retardation specifications. Finally, a framework must be used to support the ceiling panels.

    B. MATERIALS: Basic metal components required to form the frame include main runners, crossties, angle wall moldings, tie wires, and screw eyes. See Figure 6–7. The acoustical panels can be either 2′ × 2′ or 2′ × 4′ and prepainted one side. Lighting fixtures are either to fit within a panel area or are attached below and to metal runners.

Figure 6-7: Metal Framing for Suspended Ceilings **Courtesy of Armstrong World Industries**

C. SOUNDPROOFING AND FIRE CONTROL: Acoustical ceiling tiles and panels have been manufactured for many years. They are made of lightweight fibrous material and are specially engineered to be sound absorbent. Tiny holes and crevices in them trap more of the sound striking the ceiling as compared to a flat solid surface which reflects much sound back into a room. Not all types of panels are fire retardant, however, so care must be taken to obtain those that are fire rated.

D. INSTALLATION: Usually a suspended ceiling is installed after the walls are completed. See Figure 6–8. The finished ceiling height is determined from the building plans and is established on the wall by positioning a level line at this height around the room. The angle wall molding is attached above and even with the level line. Main runners are installed on 48″ oc and hang from wires attached to the ceiling joists. Each runner must be leveled during its installation to maintain a flat ceiling appearance. Crossties are installed at right angles and 24″ oc between runners. Any lighting

Figure 6-8: Suspended Ceiling Installation **Courtesy of Armstrong World Industries**

fixtures are installed before the tiles or ceiling panels are inserted into the grids formed by the crisscrossing of runners and ties. Insulation batting is installed above the tiles or ceiling panels as the work proceeds.

SYSTEM 2: A gypsum minimum 1-hour fire rated ceiling applied directly to wood joists, with sound protection (STC 40 or better) where there is a second floor.

A. Standard: Fire rated gypsum drywall panels in one of three possible arrangements, single-layer, double-layer, or resilient channel between joists and gypsum panels is appropriate. Sound and fire ratings are improved with the inclusion of acoustical, fire rated blankets.

B. Possible solutions:

1. Attic; 1-hour fire rating: ½" fire rated gypsum drywall — joists on 16" o.c. — panels attached with 5d cement coated nails 6" o.c. — joints finished. No STC rating.

2. 2nd floor; 1-hour estimated fire rating: ½" fire rated gypsum drywall — 1¼" nominal wood sub- and finished-floor — 2 × 10 wood joists 16" o.c. — RC-1 channel screw attached to joists — panels attached with 1" Type S screws — joints finished — 44 oz. carpet and 40 oz. pad atop floor. STC 47. With ⅝" drywall STC 48.

3. 2nd floor; 1½-hour fire rating: Resilient ceiling — 2 layers ½" fire rated gypsum drywall — 1" nominal sub- and finished- floor — 2 × 10 wood joists 16" o.c. — RC-1 channels spaced 24" o.c. screw-attached over base layer panels — face layer screw attached to channel 12" o.c. — joints finished. Not STC rated. Increase to 2-hour fire rating by using 2 layers 5/8" gypsum drywall. Inclusion of a 3" sound attenuation fire rated blanket install between joists will greatly improve airborne and impart sound ratings.[6]

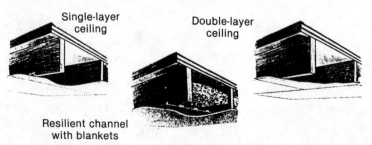

Single-layer ceiling

Double-layer ceiling

Resilient channel with blankets

Figure 6-9: Ceiling Systems for Wood Framed Ceilings, **Courtesy of U. S. Gypsum Company**

[6]USG drywall/wood frame systems/SA-924; United States Gypsum Company, Chicago, IL, 1987, pages 6 and 7.

Although it should be evident from each description how to assemble the various products, we shall briefly identify the techniques to avoid any misunderstanding.

Single layer drywall needs to be fastened so that the panels run perpendicular to the run of the ceiling joists. Ends of panels need to be nailed into the joist at the end of the panel. This may require trimming the panel so that half the joist width is exposed for the end of the next panel. These ceiling panels need to be installed before the side wall panels are installed. Nailing requirements must be met at 6″ o.c. across the panel. (Figure 6-9).

Where the design calls for a double layer of gypsum panels, the second layer must run the same direction as the first. But, the joints need to be offset 10″. The second (face) layer is bonded to the base layer with joint compound to the entire back of the face panel with beads ⅜″ wide and spaced 1½″ to 2″ o.c. Use one of three holding techniques. One, apply pressure until bonding is set. Two, use double-headed nails to hold panel. Remove nails after bonding material has set; dimple holes and fill. Three, nail the face panel through the base panel into the joists. (Figure 6-9.)

Where resilient channels are used, one technique is to apply them directly to the ceiling joists (Figure 6-10). After these are applied, the gypsum panels are attached to the channels with screws.

Where resilient channels are used in a double layer system, the base layer is applied first to the joists. Next, the RC-1 channels are installed perpendicular to the joists and screwed in place. Then the face panel is applied to the RC-1 channels.[7]

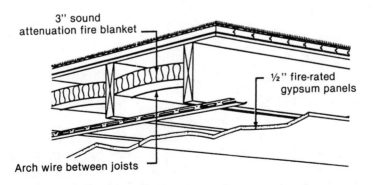

3″ sound attenuation fire blanket

½″ fire-rated gypsum panels

Arch wire between joists

Figure 6-10: Ceiling System for Steel Framed Ceilings, **Courtesy of U. S. Gypsum Company**

[7]USG drywall/wood frame systems/SA-924; United States Gypsum Company, Chicago, IL, 1987, pages 10, 11, and 12.

Conclusions

The different ceiling systems just analyzed qualify for a minimum one-hour fire rating and for sound control. Some were examined as a single-level structural ceiling, and some for multi-level structures.

The fire rated gypsum wallboard and the acoustical, fire rated blankets are essential for fire and sound control. For standard thermal protection the glass or wool blanket insulation has to be a part of any installation.

Sound control is rated by level penetration. For instance, noise created by the rain on roofs must not penetrate. In a multi-level structure walking and impact noises must also be controlled. In addition, some reflective sound control is usually a consideration. Blown ceilings offer a fair degree of sound control because of their irregular surfaces. The more irregular the ceiling surface the more the soundwaves are absorbed and diverted instead of being reflected. Finally, the acoustic tiles or panels of the suspended ceiling are most effective since their holes and crevices are designed to absorb sound. Relatively little flat surface is left for sound bounce or reflection in comparison with smooth painted ceilings.

Installation techniques are particular to each type of system. Carpentry skills are needed for all types of ceilings. The exacting installation of the suspended ceiling which must be installed level and with the frame absolutely square is a skill to be learned. The gypsum ceiling installations are varied and sometimes done entirely by drywallers. A first-class taping and floating job must be done on all joints, however, if the blown ceiling is to look good.

Door Units for Commercial Construction

The information just presented on walls and ceilings illustrates the advantages contractors gain from using various pre-finished systems and materials. Pre-finished door units also provide several advantages which we will consider later. First, however, we will define several types of standard units that are highly desirable. Then we shall examine the installation techniques used by professional carpenters for most door units.

Types of Door Units

Recall that the commercial structures studied in this text are townhouses, small office buildings, and retail stores. So many of the exterior and interior door units used in residential construction can also be used in such commercial construction.

Exterior Door Units: Figure 6–11 illustrates the typical exterior door unit. Various elements required for a unit of this type are labeled in the cutaway details. Figure 6–12 illustrates a type of complete metal door unit usually installed by store front specialists. Carpenters may need to prepare the base for the installation.

Interior Door Units: Figure 6–13 illustrates the standard interior door unit. This unit is manufactured to the specifications shown and is usually purchased unfinished. Be aware, however, that some types are manufactured for stained finishes and some for painted finishes. Those to be stained are made from clear stock and have no finger joints. Units made to be painted have finger joint jambs and casing.

Several manufacturers also offer pre-finished door and frame combinations. These units may be purchased with either 1⅜″ or 1¾″ thick doors. Various wood grained and textured vinyl materials are applied to the door, casing, and jamb. Figure 6–14 shows a cross section view of one such door unit. Units of this type meet the commercial specification Standard 236-66 and the Industry Standard 1-69. This means that the door cores are honey combed and their frames, jambs, and casings are made from particleboard with a rating of type *1* density *C* and class 1. Wood grain or textured finishes are protected with a vinyl coating.

Installation Techniques

Modular door construction has led many carpenters to the practice of installing the units without proper bracing. The "I can throw a unit in in four minutes flat" approach identifies a poor installation method, of course. Some contractors condone and even encourage this practice. Some may even have a notion that on a commercial job this results in additional profits through reduced labor costs since *"who's to know the difference"* if quality is skimped.

The truth is that the average entrance door in a commercial structure is in use much more than one in a private residence. This constant opening and closing obviously advances the wear rate much sooner than for a house. It is necessary, therefore, that the unit be installed with a sound and secure technique.

Exterior door unit

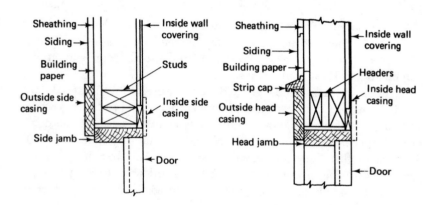

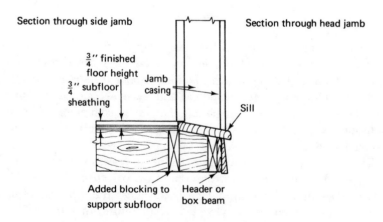

Figure 6-11: Installation Layouts for Exterior Door Unit

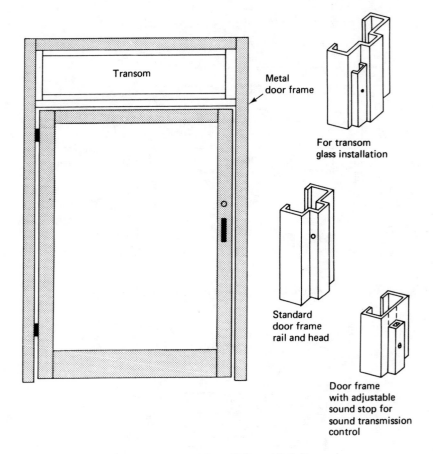

Transom

Metal
door frame

For transom
glass installation

Standard
door frame
rail and head

Door frame
with adjustable
sound stop for
sound transmission
control

Figure 6-12: Metal Door Unit and Details

Exterior Door Units: The installation of the exterior door frame
was studied in Chapter 5. The installation methods used for exterior
doors in single-wall and double-wall systems were explained in detail.
Requirements for sill preparation over a wood joist floor and for sill
removal for over-concrete floor installation were also covered.

What is of importance about exterior door installation for the
completion of the interior is the total width requirements for the jamb.
The methods for installing wall materials around these doors are also
significant.

The standard configuration of wall covering in relation to door
jamb completion is shown in the diagram at the top of Figure 6–15.
Notice that this installation requires the close fitting of paneling,

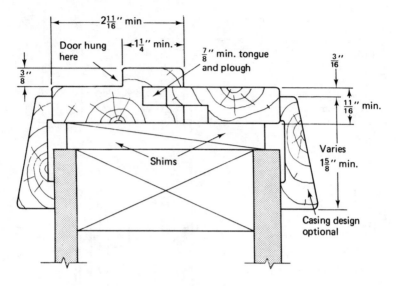

Figure 6-13: Interior Door Unit—Layout Details

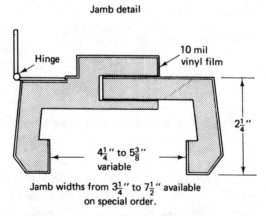

Figure 6-14: Cross-section of Vinyl-Covered Door Jamb

gypsum, and/or wood paneling so that their combined thickness is exactly flush with the inside edge surface of the jamb. The trim covers the joint and completes the job.

The center diagram in Figure 6–15 illustrates an installation where the jamb is flush with the studding and a metal or plastic edge molding is placed over the gypsum paneling. The cutting and fitting of the

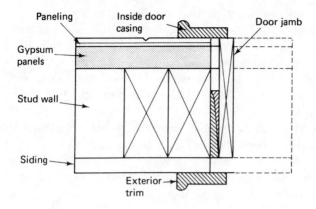

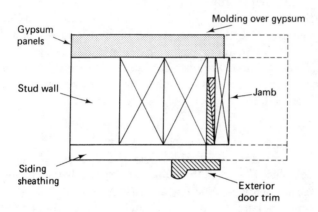

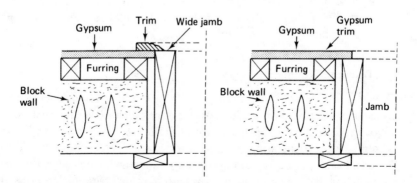

Figure 6-15: Installation Layouts for Exterior Door Frames

gypsum panels must be exact so that the trim laid over the edge of the gypsum panel can be aligned parallel with the inside edge of the jamb. The amount of offset is from ¼″ to ½″ as a rule. This second method is usually employed where pre-finished gypsum panels are installed, although it can successfully be used with unfinished panels. Finally, note that the jamb width is different from that shown in the diagram at the top of Figure 6–15.

Either method could be used in finishing an exterior door unit installation for either a concrete block or a poured concrete wall. The jamb's width would need to be wide enough to allow for furring and gypsum panels as seen in the two diagrams, bottom of Figure 6–15.

In conclusion, a word about savings. The second method which uses gypsum trim would eliminate the cost of standard wood or vinyl trim as well as its installation time.

Interior Door Units: Interior door units are separated and the half with the door is installed. Bracing between jamb and stud must be installed. Then the remaining partial jamb and its casing is installed. A sound interior door installation requires the same attention to detail as for exterior door installation. The following steps constitute a quality procedure:

1. Trim either side jamb at the floor level to make the head level and to be sure both jamb sides rest on the floor.
2. Plumb and straighten jamb sides. Use the door as a straight edge to do this. Maintain the standard spacing between door edge and jamb while face-nailing the casing to stud.
3. Install bracing between the jamb and studs at the hinges, striking plate, floor, and head areas. Nail bracing with finishing nails since part of the material is exposed.
4. Install the remaining half of the jamb after the bracing is installed. Once jamb is in place and face-nailed, nail it through the braces into the stud.

Trim and Trimming

Once again study of a subphase of interior work on commerical structures can start with a review of residential construction. All techniques and many of the materials used in completion of residential trimming operations apply to trim work in commercial building. They include installation of window stools and aprons, boxing or self-return of paneling for completion of window framing, trim for interior of exterior doors, of installation of ceiling molding, and of installation of baseboard and shoe.

We shall see with the aid of the chart in Figure 6–16 that trim may not always be necessary. Its use depends upon the combination of wall and ceiling systems installed for the interior of a structure. In addition to its utilitarian purpose of covering joints, trim is frequently selected to enhance structural beauty, create design moods, break up too-large areas visually, or shift dimensional emphasis. These matters are *not* discussed in this study since their use is so varied and is largely a matter of personal taste. But, if *special* effects are to be incorporated in a construction project, these requirements must be carefully integrated with the standard trimming operations.

Types of Trim

Before we venture too far into this section, let's identify the typical trim that can be used on commercial structures. Figure 6–17 shows examples of the common mill types of trim. Those illustrated are made from wood and may be purchased pre-finished. Such pre-finished products are available in a variety of wood tones and grain patterns and the surface finish usually has some type of vinyl coating.

Cellular vinyl (plastic) moldings are also available to the building contractor. See Figure 6–18. Those shown are finished in wood tones and in black, white, and gray; the colors are permanent. The material is resistant to water, humidity, and physical abuse. These characteristics are desirable for commercial application.

General Requirements for Trim

Townhouses and small office buildings require essentially the same trimming operation. Windows and the inside of exterior door jambs need finish trim. Baseboards will be installed and, in some cases, shoe molding. There may be a need for ceiling coves. Some limited special effects and custom work may be wanted, say on stair wells, stair cases, and decorative ceiling beams.

Retail store and restaurant trim may be limited to windows and baseboards. Often the doors are aluminum and glass and do not require trim. But again some special effects may be required. Usually from general trim requirements we can proceed to the more specific.

Combination of Wall and Ceiling Systems

If a wall system is combined with a ceiling system, specific requirements for trim and trimming operations are easily defined. Figure 6–16 refers to four examples of combination wall and ceiling finishes.

Ref. no.	Wall type	Ceiling type	Ceiling trim B	Baseboard A	Baseboard B	Shoe A, B	Window stool and apron	Window casing A	Window casing B	Window self-return C	Ext. door casing A, B
1	Gypsum	Gypsum (blown or painted)	N	Y	Y	Y	N	N	Y	N	Y
2	Gypsum pre-finished	Suspended acoustical	N	Y	N	Y	N	N	Y	N	Y
3	Pre-finished paneling	Gypsum (blown)	Y	N	Y	Y	Y	Y	Y	Y	Y
4	Pre-finished paneling	Suspended ceiling	N	Y	N	Y	N	N	Y	Y	Y

A: Unfinished trim stock
B: Pre-finished trim stock
C: Wall material is used for returns
Note: Y = yes   N = no

Figure 6-16: Trimming Requirements Chart

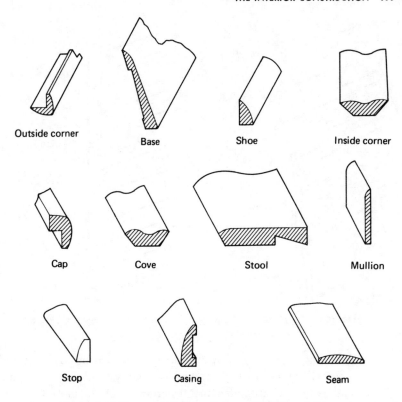

Outside corner   Base   Shoe   Inside corner

Cap   Cove   Stool   Mullion

Stop   Casing   Seam

Figure 6-17: Types of Wooden Door and Window Trim

By using the paired wall and ceiling systems as constants the trim items become the variables. The variables include ceiling trim (cove, crown) baseboard, shoe, window trims and casings. Notice that there are three footnotes in the chart. *Footnote A* indicates that unfinished mill stock trim in its natural state is to be used. When this is the case, plans for its quality, texture, and finish must be calculated as added work. *Footnote B* indicates pre-finished stock which can be either pre-finished wood stock or pre-finished cellular vinyl molding. *Footnote C* indicates that the wall material *may* be used for the window boxing or self-return or that stock materials such as lumber or plywood may be used. Now let us see what the chart shows.

1. As indicated in Combination 1, the wall is to be enclosed with gypsum paneling with taped and floated joints. The ceiling calls for gypsum paneling with taped and floated joints and can be either a blown or painted finish. Looking across the chart we

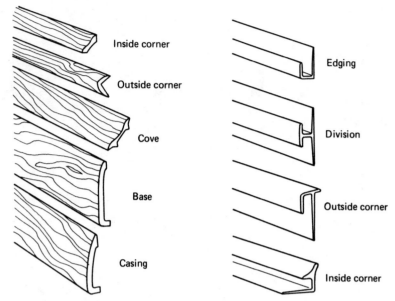

Figure 6-18: Types of Cellular (Vinyl) Trim for Windows and Doors

find that *no* ceiling trim and *no* window casing are required. The self-return on the window is usually made from gypsum and is floated around a metal corner bead. In most instances the trim can be limited to a sill or stool and apron. Notice that baseboard, shoe, and door casings are needed. Again, remember that the trim materials are the variables.

2. In Combination 2, the wall is to be covered with pre-finished gypsum panels. The ceiling is to be the suspended type. Usually such a ceiling is installed after the walls are completed. The wall channel for the acoustical tiles will fasten through the gypsum panel to the studs. This arrangement eliminates the need for ceiling molding. Note that pre-finished baseboard and casing trim are recommended. Also note that the windows are to be self-returned with pre-finished gypsum paneling.

3. In Combination 3, the wall is to be enclosed with an outer layer of pre-finished paneling. The base layer could be gypsum wallboard and/or sound deadening board. The ceiling is to be of gypsum wallboard and taped, floated and blown. This means that the ceiling has to be completed *before* the paneling is installed. Checking now in the chart we see that there is a need for ceiling trim. Pre-finished trim stock is recommended for the moldings since it matches the paneling and eliminates the

chance for painter error. You can expect to perform all the other operations. The accepted finish method for the windows is to use self-returns made from paneling pieces.

4. In Combination 4, the wall is to be enclosed as in Combination 3. With this combination, however, the paneling is usually installed before the ceiling. The ceiling's wall channel is nailed onto the face of the paneling. This eliminates the need for ceiling trim. Once again pre-finished trim stock, either wood or vinyl, should be used to complete all other operations.

Conclusions

There are several conclusions that can be reached from this study of finish work. First, the kinds of combination of wall-ceiling finish dictate the sequence for installing the materials. This in turn has a direct bearing upon whether or not ceiling trim is needed. Secondly, the type of wall finish directly affects the selection of trim to be used. For example, unfinished gypsum paneling may require using natural wood stock trim. In contrast, pre-finished paneling and face-layer gypsum panels lend themselves to the use of pre-finished trim stock. Thirdly, the best materials options available to the contractor should be selected wherever costs for materials can more than offset the cost for highly skilled labor. Remember, work with pre-finished stock trim or vinyl is complete when it is installed.

Installation Techniques

If unfinished mill stock trim is used the standard application techniques include mitering, coping, and butting of joints. Finish and casing nails are used, and each nail is set below surface. Glued joints are also recommended for quality workmanship.

If pre-finished trim stock is used, like standard mill stock, it must be cut to fit. Coping, mitering and butting cuts are common and frequently done in the miter box. Pre-finished trim is often nailed in place with pre-painted nails that match the color of the trim. Alternately, the trim can be glued in place. Gluing techniques eliminate the use of nails. Sometimes it proves useful to combine nailing and gluing techniques.

Delays Incurred and Clean-up Work

In every construction job, whether residential or commercial, certain hidden costs are built in, as it were. They result from the inevitable delays and from the clean-up work. The delays are usually incurred during completion of one or the other subphase of the interior carpentry. For example, carpenters who install the suspended ceiling frame are often delayed by the electricians and the heating workmen. Light fixtures must be set in place, connected, and tested and heating vents and return ducts must be fitted in before the carpenters can install the ceiling panels. There is usually another short delay in the trim work if shoe molding is to be used since the flooring crew must install the flooring before the carpenter can go ahead and install the shoe. There are other similar instances in almost every phase of the work.

The clean-up work at the end of the interior phase may take from a day to several weeks, depending on how orderly or disorderly the work was. Installing hardware, replacing any defective, marred, or damaged materials, re-tightening trim joints, and checking out mechanical devices are among the many chores that are usually done at this time.

Some clean-up work cannot be avoided, of course, but most of it is caused by poor or careless work. Remember, good training methods can reduce if not eliminate both the unreasonable delays and unnecessary clean-up work. Be sure to account for this time and for some delays from work overlap in your plans and estimates since they represent hidden costs that don't stay hidden.

TRAINING ON INTERIORS

A variety of skills are used to install interior materials; they range from the very limited to the highly specialized. Personnel who are assigned to installing wall and ceiling systems need to understand clearly what they must do and why they must do the job. The tasks themselves do not require a great deal of skill. Gypsum panels, resilient channels, insulation batts and sound deadening boards must be measured, cut, applied. The tasks are repetitive and easily accomplished.

The installation of pre-finished gypsum paneling and the associated trim is a more exacting task and requires much more skill. Extensive training is required for those new to the sequence of the application and to the variety of materials used. Such work is classed as *finish* and is assigned to experienced craftsmen ordinarily.

The installation of paneling, wood or mineral, is very demanding work. All joints and cuts must be accurate and quality tolerances impose very strict limits. Usually on-job training is lengthy and therefore expensive to a contractor, but errors made by apprentices or the unqualified result in even more expensive waste and marred, unacceptable productions.

Trimming operations are very technical and require considerable skill. Some of the techniques used by qualified craftsmen can be learned by apprentice carpenters by the OJT (on-job-training) method.

Moreover, short group training sessions can anticipate some problems and will prepare the work crews for expected delays. They can also be schooled in ways to give the support that the subcontractors often need. Support skills can run from the special and limited to all-around top qualifications and are usually expected of the senior craftsman in the group.

Recording the Training

Prior to the start of interior completions an identification of available skills must be made in order to plan the work well. Each training record should be annotated as the employee is interviewed. When all employees who are expected to participate in the interior work have been evaluated it is possible to define the training requirements. See Figure 6–19 on page 204. Those who require additional knowledge of tasks should be assigned home study. They can also be paired with the highly skilled workmen for those important tasks which, if well done, improve the capabilities of the less skilled. Such a method also results in on-job training with very little overall loss in productivity.

Task Activities

Completing the interior of a structure more or less involves six major task activities. Not all six are always used on each job, but rather some combination of several. Each task activity detailed next is a training device. When well used, they will all help establish basic and full understanding of the interior work subphase.

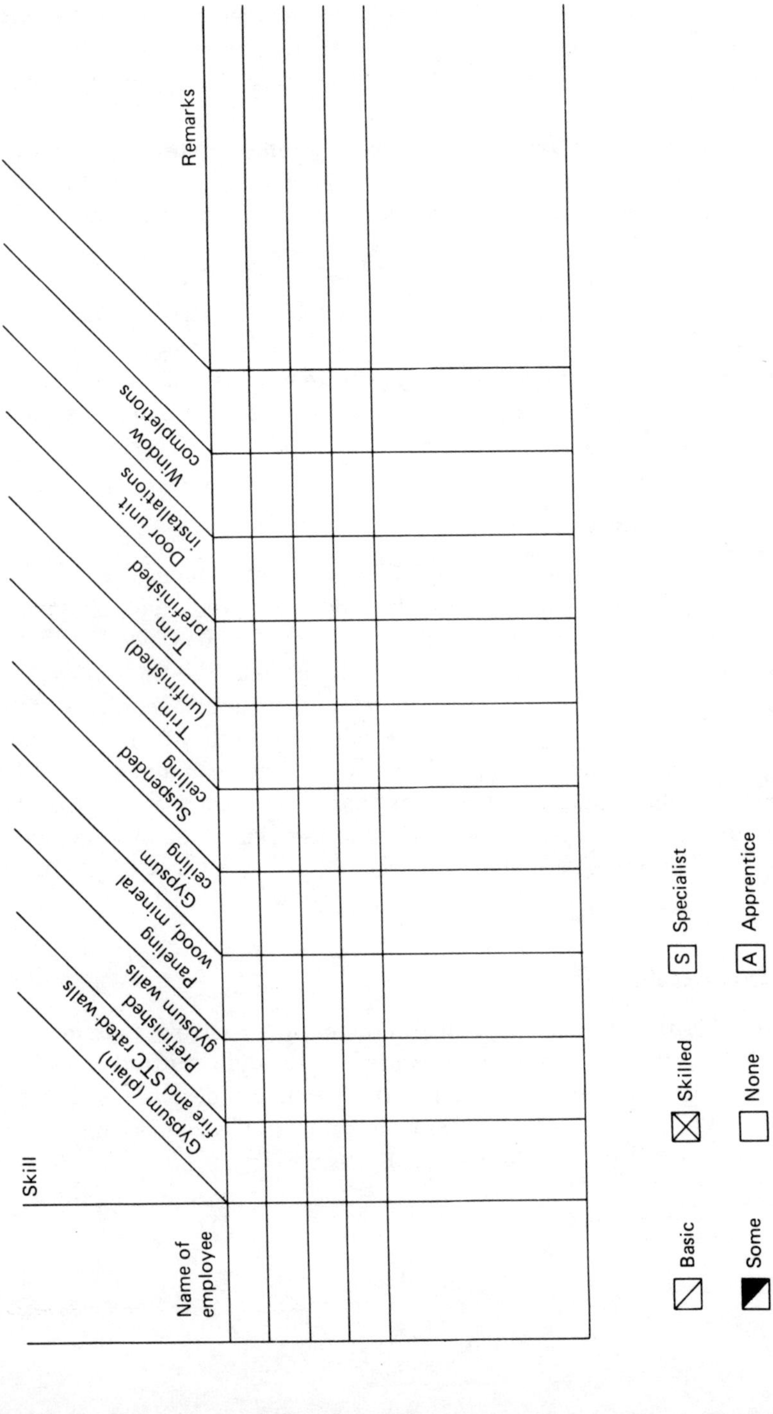

Figure 6-19: Training Record for Interior Construction Crew

INT 1 TASK ACTIVITY: PLAIN GYPSUM WALLBOARD FOR WALLS AND CEILINGS

RESOURCES

Estimated Manhours for 2 men: 1.6 to 2 mins per sq ft

Materials:

resilient channels
nails or screws
glue (if required)
gypsum wallboard
sound-deadening boards

Tools:

hammer	electric drill with screwdriver bit
straightedge	folding ruler
utility knife	T-square
	chalk line

PROCEDURES

A. WALLS

1. Lay out 24" oc horizontal spacing for resilient channels; measure, mark, cut, and install channels. Measure, mark and pre-cut first base panel outward from corner; install the panel with screws. Measure, mark, pre-cut, and install succeeding panels until entire wall is covered.

2. Lay out face panel to be installed perpendicular to base panel; cut to fit. Install with screws ensuring a dimple in each intermediate stud; screw or apply glue to rear of panel and put in place. Measure, mark, cut, and install succeeding panels until wall is complete.

3. Repeat the process outlined in Steps 1 and 2 for all walls.

4. Cut out window, door, and utility openings as each panel is being installed. On outer corners install metal corner strips as required.

5. Complete wall installation by applying sealant around all perimeters; tape joints and spackle screw heads.

B. CEILINGS

1. Prior to installing resilient channels, inspect framing to determine adequacy of securing surface (for nails or screws) meant for channels and gypsum panels. Correct any deficiencies found.

2. Lay out 24″ oc placement for resilient channels; install by whatever means required (nail, screw, clamp or wire).

3. Measure, mark, and pre-cut first base panel. Erect appropriate scaffold and install panel. Measure, mark, cut, and install succeeding panels. Ensure proper spacing of all fasteners.

4. If face panels are to be installed, lay out first one for perpendicular run. Measure, mark, cut and install panel. Proceed across the ceiling using the same procedure as outlined.

5. Tape all joints and spackle all intermediate joist fasteners; float ceiling as required.

INT 2 TASK ACTIVITY: PRE-FINISHED GYPSUM WALLBOARD FOR WALLS

RESOURCES

Estimated Manhours for 1 man: 15 mins per panel (avg.)

Materials:

pre-finished gypsum panels
pre-finished trim, runners, moldings
pre-finished nails, screws
sealant
adhesive

Tools:

utility knife
folding ruler
level
hammer or plastic-headed hammer
chalk line
straightedge

PROCEDURES

1. Stand panels (vertically) along wall and arrange so that colors and tones blend to create a pleasing effect. Number panels so that they can be installed in order.

2. Install ceiling and floor runners if to be used. Install inside metal corner strips before inserting first and final pieces of paneling into corners.

3. Measure, mark, cut, and install panels. Either butt joints or insert a divider strip between vertical joints. Finish edges around doors and windows with pre-finished metal or vinyl trims. Glue and/or nail panels in place.

INT 3 TASK ACTIVITY: PANELING FOR WALLS

RESOURCES

Estimated Manhours for 1 man: 15 mins per sheet (avg.)

Materials:

paneling	colored nails
pre-finished paneling trim	adhesive

Tools:

fine-tooth handsaw	brace and auger
straightedge	block plane
folding ruler	framing square
chalk line	hammer

PROCEDURES

1. Position each panel to determine the best layout with minimum cutting. (This step presupposes that any and all base wall panels are installed.)

2. Measure, mark, and, cut panel. Try it for fit in its intended place; remove and apply adhesive to base as required. Re-install and press firmly to base covering or to studs if base is not used. Nail at ceiling and floor levels to hold panel while glue sets.

3. Repeat Step 2 for each succeeding panel. If division strip is used, insert it between vertical joints of panels.

4. Scribe pieces of paneling to fit accurately into corners, unless inside corner strip (metal or vinyl) is used. Cut pieces very close to the line to ensure an accurate fit.

5. If nails are used, drive pre-painted matching nails through panels into studs behind panels.

INT 4 TASK ACTIVITY: DOOR UNIT INSTALLATION

RESOURCES

Estimated Manhours for 1 man: interior 15 to 30 mins (avg.)
exterior 1 to 2 hrs (avg.)

Materials:

interior door unit   casing nails
blocking material   finish nails

Tools:

hammer   handsaw
level   chisel
folding ruler

PROCEDURES

1. Separate two halves of door jamb unit.

2. Insert the half with door into the opening and tack-nail casing to the studs while shifting the jamb to make all spacing between door and door jamb equal and uniform.

3. Insert blocking between jamb and stud openings behind the hinge, the striking plate areas, and at the head and floor levels on sides of jamb; nail in place with finish nails.

4. Insert the remaining section of jamb and trim; face-nail to studs.

INT 5 TASK ACTIVITY: WINDOW COMPLETIONS

RESOURCES

Estimated Manhours for 1 man: 20 mins to 1 hr (avg.)

Materials:

wall-stock material
corner beading
window stool or sill
apron material
blocking material

Tools:

level	snips
combination square	hammer
folding ruler	hacksaw
utility knife	handsaw

PROCEDURES

A. GYPSUM AND PANELING RETURN (self-return)

1. Install backing at sill, center, and head areas along window sides to *fur out* the area so that when the wall material is installed its outer surface is at required level.
2. Repeat Step 1 for the window head area.
3. Measure, mark, cut, and install window stool or sill. Allow sill to extend past window opening ¾ inches each side and 1¼ inches into the room. (*Note:* If casing is to be used, add ¾" end overhang to width of casing.)
4. Measure, mark, cut, and install wall material into the head area of.the window. Repeat process; install side pieces.
5. Add metal corner bead for floating.
6. Add outside corner bead for pre-finished gypsum or paneling (wood or mineral).

B. STOCK LUMBER RETURN

1. Calculate thickness of blocking material needed to bring finished stock lumber to desired point on metal window frame; cut and install pieces at head, center, and sill areas.

2. Repeat the process for the sill area.
3. Measure, mark, cut, and fit sill into place, allowing for width of casing plus overhang and 1¼" into room; install sill. Cut and install apron below the sill.
4. Measure, mark, and cut the head piece; try for fit; correct as required and install.
5. Measure, mark, cut, and install side pieces.
6. Trim outer edge with casing; using a butt cut at sill area and miter cut at head area. Glue mitered pieces as they are being installed.

INT 6 TASK ACTIVITY: TRIMMING OPERATIONS

RESOURCES

Estimated Manhours for 1 man: 20 mins per door casing
10 mins per 12 ft of baseboard
10 mins per 16 ft of shoe
15 mins per 12 ft of crown or cove

Materials:

casing
shoe
nails, finished and pre-painted
baseboard
crown or cove
adhesive

Tools:

fine-point handsaw
hammer
coping saw
miter box
chisel
folding ruler
block plane

PROCEDURES

1. Measure, mark, and cut casing side-jamb piece and install on door jamb. Measure, mark, and cut miter on head piece of casing to match first installed piece. Try for fit then mark the opposite end; cut it on a miter and set piece aside. Cut and fit the opposite side-jamb casing. Position and tack-nail the second side-jamb casing piece in place. Try the head piece again; make adjustments as required. Nail all pieces after applying adhesive or glue between mitered pieces.

2. Install the first baseboard piece after cutting both ends with butt cuts. Cope the end of the second piece to fit at the corner; cut for length or miter if splice is needed. Continue around the room. Secure each piece of baseboard to the wall with nails or adhesive.

3. After tile flooring has been laid, install shoe; use coping cut at inside corners and 45 degree splices on long runs. Nail or glue stock as appropriate.

4. Cut the first piece of crown or cove to reach from corner to corner; install with nails. Miter and cope the next piece of molding; try for fit; cut for length. On long walls, miter all splices; continue cutting, fitting, and installing pieces. On outside corners make the joint by mitering both the pieces that meet.

REVIEW QUESTIONS

1. Why is it essential that the production foreman participate in the planning phase of interior construction?

2. What standards must be met if a wall is to be 2-Hr rated against fire with *STC 50* or better?

3. Make a comparison chart of 1-Hr rated walls of wood and of steel studs; define the differences.

4. What is the mineral added to gypsum that aids in fire retardation?

5. If certain walls are to be constructed with different requirements how can they affect contract costs?

6. Name two advantages for the contractor when he uses prefinished wall paneling.

7. How can the use of metal and vinyl stripping ease and speed up the installation of pre-finished paneling?

8. Why is a suspended ceiling usually installed after the walls are installed?

9. Why is gypsum paneling installed on a ceiling prior to installation of a wood ceiling?

10. Which type of ceiling offers the best sound transmission and reflection control?

11. List the major steps a carpenter must take when installing an interior door unit.

12. Explain one of the three methods for finishing the interior of an exterior door jamb.

13. What trimming requirements are common when a suspended ceiling and paneled wall are combined?

7

Cabinet Installation and Cabinetmaking

Installer: a skilled workman who fits and puts cabinets in position and ready for use; also said of anyone who puts equipment in service
Level: the act of measuring a flat, even surface or of correcting so as to obtain a surface condition without any irregularities (no part higher than another)
Planer: a type of saw blade that makes a fine or planed cut
Plumb: a vertical condition usually determined with the aid of a spirit level or weight (bob) hanging from a cord; a precondition of walls and floors for cabinet installation
Rabbet: a joint made along edges or on a cabinet door or drawer so as to form a recess or interior angle where another element may nest or unite; used for offset design features

Rails: the bars of wood or the like serving as horizontal members to frame or trim a cabinet
Scribe: a method for marking or scoring a board or panel with a pointed instrument or knife as a guide to cutting and fitting, especially irregular surfaces
Stiles: the various upright members of the frame or trim on panels, especially on a cabinet
Straightedge: a bar or strip of wood or metal having at least one edge of reliable straightness and used to verify straight lines, plane surfaces, and the like; in wood usually made of 1¼" stock in 6" widths and from 3'6" to 7'6" long
Wood plugs: either dowels or machined pieces used for joining two members; as in a dowelled joint

Cabinet installation and cabinetmaking tasks are given only to the most experienced craftsmen. Work as a trade journeyman does not qualify a carpenter as a cabinetmaker; it does, however, give him a foundation for learning how to do the quality work required of a cabinetmaker.

You may have noticed that there are two distinct subjects in this chapter. Being a good installer requires an understanding of cabinet structures and their installation techniques. A good working knowledge of stress factors coupled with an understanding of ordinary design characteristics usually provides the cabinet installer with the data he needs to make safe, satisfactory installations. Joinery, stress joints, dadoed shelving, and mortise and tenon frames are important construction details.

Cabinetmaking, however, involves the very skillful translation of a set of plans into a fine custom built cupboard or cabinet. It includes great familiarity with design features and experience at the expert selection of the type of joinery required. It may also include appraisal and selection of special materials needed to accomplish a given cabinet's functions.

This chapter covers the numerous activities associated with cabinet installation and gives limited coverage of cabinetmaking as well.[*]

Based upon what has already been said about prefabrication, we can see that the installation will warrant standard methods. Assorted pre-built cabinets are made for townhouse kitchens and bathrooms. There are also pre-built display cabinets for stores and shelving for both stores and offices. You must also learn data related to custom cabinetry for wall cabinet-shelving combinations, for base counters, and for display window shelving.

As for the other phases of the construction contract, the cabinetmaking installation phase requires organizational planning and control. Since special skills may be needed or subcontracts may be issued, relevant decisions need to be identified now.

ORGANIZATIONAL FUNCTIONS IN CABINET WORK PHASE

Figure 7–1 again shows four-part organizational planning: scheduling, production, skills acquisition and training, and material. All these are involved in cabinet installation and cabinetmaking. A study of the structure's plans and specifications will identify the requirements for cabinets, their types, and whether it is likely that they will be subcontracted or built on the site.

There are two separate approaches and a mixed approach to cabinet installation. The first, separate aproach, is to define the cabinet work as subcontractor-required. With such an approach the materiel man will treat any cabinet as a *long-lead item*. So he will put bids to

---

[*]A comprehensive study of cabinetmaking can be found in the *The Complete Book of Woodworking and Cabinetmaking* by B. W. Maguire (Reston, Va.: Reston Publishing, 1974).

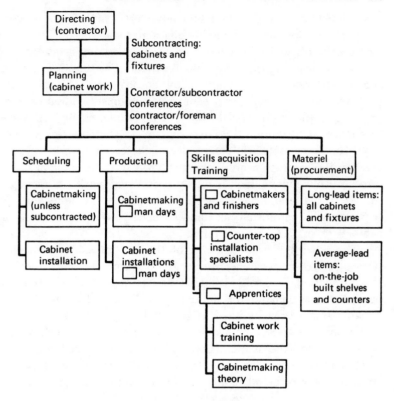

Figure 7–1 Organization of Cabinet Work Activities

various cabinet shops as early as possible. The general contractor will negotiate the necessary contracts, ensuring that the specifications are clear and resolving any conflicting requirements. His negotiations will also make the scheduled delivery dates firm. These dates if correctly assessed by scheduling will "blend" the work properly for on-time completion. The contractor's regular carpenter force will have minimal or no involvement. In effect, they will provide little if any support since the maker of the cabinets installs them.

The second approach is for the contractor work force to *build in* the cabinets as a function of the overall job. In this case, materiel will need to secure the materials which, except for unusual specifications, may be treated as *average-lead items*. Most materials, including cabinet-grade plywood, hardwood trim, hardware, and counter topping, are usually available locally. Opting for built-in cabinets means that scheduling must allow more work time for this subphase than otherwise. The work must be integrated with the other interior work as well as with the other subcontractors—painting and electrical, for instance. Special workmen's skills and some special attachments for power tools will be needed. The ripsaw blade on a table saw, for instance, will be replaced by a planer blade. For the most part, glue and wood screws replace nails for the construction. The specific skill requirement means that either cabinetmakers must be hired or trained. Training in the theory and techniques of cabinetmaking will have to be made part of the job training program if cabinetmakers are not to be hired. The variety and complexity of the cabinet work will dictate whether the contractor hires a cabinetmaker or trains some of his work force. Since cabinetmaking is highly specialized work, no assumptions should be made about even a good carpenter's cabinetmaking skills. A person who can install door and baseboard trim can not also build cabinets unless he is trained for it.

The third approach is a mix of the first two. The cabinets are subcontracted to a cabinet shop but the carpenter work force does the installation. This mix makes a long-lead item of the materials and necessitates a demand for limited specialized skills for installation. A cabinetmaker can be hired as an installer since his knowledge and skill insure a satisfactory installation. Or part of the work force can be trained for the work. They will need to understand how the cabinetmaking shop has provided for installation and what particular stress forces must be taken into account during the installation.

CABINET INSTALLATION

We start the detailed study of cabinets and cabinet installation with a four-part break down: 1) base cabinets, 2) wall cabinets, 3) display cabinets, and 4) display shelves. Each is examined from the same point of view whether partly or wholly pre-built.

Base Cabinets

A base cabinet is defined as a storage cupboard that rests on a floor and against a wall. The usual configuration is shown in Figure 7–2. Notice that it has a *toe space* so that persons working on its top and/or in front of it can stand up to the cabinet and work comfortably. The space within the cabinet may have any arrangement of doors, shelves, and drawers that provides organized storage space. Such base cabinets are usually manufactured in 34½" heights, in 22" and 22½" depths, and from 9" widths, increasing each three inches—to 12", 15", 18", etc.

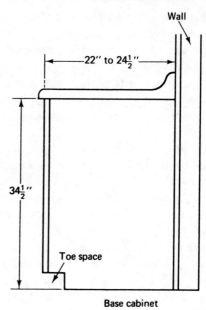

Figure 7–2 Layout for Typical Base Cabinet

The installer must follow a particular sequence to install base cabinets:

1. Survey the floor surface for level and irregularities.
2. Decide upon anchoring method.
3. Mark cabinet(s), position, and install anchors.
4. Set cabinet against or into anchors and secure.

Surveying the floor area involves the use of a level and a 6-ft straightedge. The straightedge provides a check on the long dimension of the floor surfaces and it identifies irregularities, such as fractions of inches of depressions and swells. Placed on top of the straightedge the level measures the general slope of floor. The level is also used to check the plumb accuracy of the back wall. Held horizontally the straightedge on the back wall identifies any irregularities there. Where problem areas are located they are marked with chalk. Notations must be made carefully of any fractional depressions or swells and of all out-of-level, low, high, or out of plumb surfaces as well as the direction in which the wall leans. Corrections of these conditions make reliable fitting possible.

Deciding on the anchoring method means asking two questions: "What is available to anchor to?" and "How secure must the base unit be?" In answer to the first question, it is likely that the carpenters have already installed drywall or paneling in the structure. So they either know or can find out about the wall and whether there is a wood-framed joist or concrete floor installed. Either drywall or paneling means that wood screws and nails can be used to secure the back of the cabinet to the wall framing. If the floor is wood, a block of 2×4 installed with nails is sufficient anchor. But, if the base unit is to rest on a concrete floor the 2×4 blocking can be installed by plugging and screwing or by nailing with case-hardened nails or with a stud driver.

The answer to the second question comes from answering the first—by putting in a stable anchor. Stability is essential. No movement—sideways, front to back or lifting upward—is permissible.

Eliminating surface irregularities from the area of installation and choosing anchoring methods brings us to the next step. At this point mark the position for the first and for the succeeding cabinets. Position the first cabinet temporarily, in the corner, under the window, or along the wall—however the floor or detail plan specifies. Level the cabinet, front to back and side to side, using shims if needed. Measure and cut shims as needed to fit under the base and between cabinet sides and wall. With the cabinet properly positioned the location for the wall anchors can be marked on the cabinet back. The outer perimeter of the unit should be scribed on the floor. Also scribe the base and cabinet sides for trimming if it is necessary. Remove the cabinet and install a floor block; set it back from the scribed line by the thickness of the cabinet toe plate and cabinet sides. If lag screws or the like are used as wall anchors, pre-drill holes through cross supports at the back of the cabinet to allow passage of the screws.

If shims are used, cut and nail them in place while the cabinet is away from its position along the wall. Do any needed trimming by cutting along the scribed lines on the cabinet. Then set the cabinet in place. Anchor it to the wall and floor, using screws and/or nails. Repeat the process until all cabinets are installed.

The next operation is to install the counter top. It is presupposed that the counter top is a unit or set of pieces forming a unit. There are four key points to a proper installation:

1. A survey that ensures correct fitting of tops to cabinets.
2. Accurate identification of fastening areas; use of good anchoring method.
3. Making joints true and securing them properly.
4. Assembling tops properly on the cabinet.

Pre-formed counter tops are normally used by the installers in commercial construction. On occasion, the carpenters may be required to prepare a plywood base for counter top installation specialists. Many of the aspects of the work discussed next would apply to such a situation; however, this study concentrates primarily on the installation of pre-formed counter tops.

In all likelihood the cabinetmakers in the cabinet shop that supplies the cabinets for the job will also prepare the counter top assemblies. They will, from the plan furnished, prepare the needed pieces of measured and molded or rolled topping. This means that your task as a carpenter should be relatively simple if exacting.

Before you install the top, place the straightedge across the previously installed base cabinets. If the installation was accurate there should be no irregularities. While you are making this survey, also place a level across the straightedge and verify the levelness of the cabinets. Identify any irregularities and take steps to correct them. Also, check for level from front to back on the cabinets.

Next, decide where the fasteners (screws) can be installed from inside the cabinet into the counter top. If no blocks are provided, you must now install them. There should be some blocks at all the corners and at regular intervals around the edge of the top of the cabinet. If glue is used along with the screws, sand the gluing surface lightly wherever it is rough to ensure better bonding.

Place the cabinet top pieces in position and check the joints. These must be completely accurate. If the wall's corner is not made at exactly 90 degrees, one of two things must be decided: either to recut the miter on the cabinet top to allow the back of the counter top to seat against the wall, or to place a molding on top of the counter top to fill the space between counter top and wall. Of the two decisions, the first is better. It does present problems however, since both halves of the miter need to be recut. The supplier who delivered the top is better able to re-do the cutting. The second, poorer solution requires unplanned expenditures of time and material and, therefore, should be avoided if possible.

It is customary to prepare joints in counter tops for clamping. This process involves grooving slots in the underside of the top as Figure 7-3 shows. Clamps are fitted into the slots and tightened. During this operation glue is inserted in the joint.

Attaching the counter top to the base unit involves, as we have seen, the assembly of sections of tops and the fastening of complete tops to base units. If possible, the clamping task should be performed first, then the screwing and gluing of top to base unit should follow.

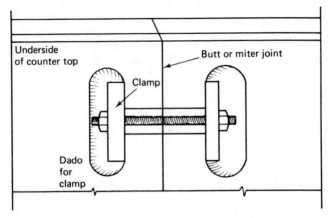

Figure 7–3 Method for Clamping Counter Tops

This procedure eliminates difficulties that otherwise might occur if sections are clamped with cabinet partitions, drawer guides and blind corners in the way.

These are the standard principles for base cabinet installations. Almost without exception each installer also has to make certain on-site modifications regarding these principles. His sound judgement results in a satisfactory installation.

Wall Cabinets

But what of the wall cabinets and their installation? There are two types of wall cabinets that can be installed by the interior finish crew. One is used over base cabinets in townhouses and restaurant kitchens. The other is the floor to ceiling type that has racks for hanging clothes or adjustable shelves for displaying merchandise. The techniques for installing each type is slightly different. Each type, therefore, is discussed separately.

The wall cabinet shown in Figure 7–4 is subject to dead weight and pull stresses when filled. This stress is mainly downward and forward. So installation techniques must combine to work with the design characteristics (see page 221) to keep the cabinet in place. Some of the installation principles for base units are applicable to wall units. The installation principles for wall cabinets placed over base cabinets include the following:

1. Selecting the proper method.
2. Sounding and installing the anchors.

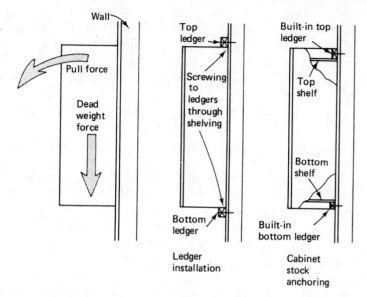

Wall

Pull force

Dead weight force

Top ledger

Screwing to ledgers through shelving

Bottom ledger

Ledger installation

Built-in top ledger

Top shelf

Bottom shelf

Built-in bottom ledger

Cabinet stock anchoring

Figure 7—4 Anchoring Methods for Two Types of Wall Cabinets

3. Securing the cabinet to the wall.
4. Blending the cabinet(s) with "drop" or false ceilings.

Two standard methods are used to install such wall units. One requires the installation of a ledger board upon which the cabinet sets; the other depends on cabinets that incorporate the use of built-in pieces of stock which are inside and along the cabinet's back surface. These built-in ledgers are integrated in the cabinet construction so as to take fasteners that hold the cabinet to the wall. Either method usually results in a satisfactory installation. The advantage of using a ledger board on the wall is that it is easy to handle and install. The cabinet is then set onto it and secured to the wall. The advantage of using the cabinet with the anchor stock built into it is that it is always available; no additional material is needed. A cabinet's weight and bulk must be taken into account, however. Ledger board can usually take more weight than most cabinet stocks.

Secure foundations for the anchors, such as wall studs, are not always where they are needed. The "sounding" technique is frequently used to find out just where they are. Gentle tapping of the wall obtains the dull "thud" sound that identifies the stud position. From that point other studs are either 16″ or 24″ oc. The studs make an excellent anchor for the ledgers.

Anchors, such as mollies or wood plugs have to be installed in a concrete wall. Such a wall is an excellent place to use the ledger-below cabinet method. Anchors can be installed in the wall by using star drills or an electric drill with mason bit. It is also desirable to install a ledger strip along the line of the top of the wall cabinets if space permits.

Two pieces so located allow workmen to secure the cabinets from inside into the ledgers. Pre-drilling holes through shelving aids in the installation. Then while one person holds the cabinet in position, another inserts the screws.

If a cabinet with built-in anchor stock is used for anchoring purposes in lieu of wall-mounted ledgers, pre-drilling for screws should be done. In this type of installation, transcription of stud or anchor positions must be made from wall to cabinet. As each is installed it must be carefully aligned along a predetermined height line.

Securing means not only fastening the cabinet to the back wall but cabinet to cabinet. So facings must be perfectly aligned, then screwed together. Since they are made from hardwood, pre-drilling is essential. It keeps the stock from splitting and prevents improper aligning of the screw.

Blending both the individual cabinet or a group of cabinets to a false or drop ceiling requires that one of two methods be used. With the first method, the framing and covering of the drop ceiling, whether paneling or drywall, are designed to protrude into the room several inches farther than the cabinet facing. See the left detail in Figure 7–5. If this is the design, a shoe or cove molding is usually installed between the cabinet top and the ceiling.

With the second method, the paneling or facing gypsum is either set flush or to just overlap the cabinet top rail. The framing may be installed either before the cabinets are set in place or after. The essential requirement is that the material from cabinet top to uppermost ceiling area be plumb. If the ceiling framing and covering are installed before the cabinets, great care must be taken to develop the finish lines of the vertical materials so that they blend with the cabinet facings.

Wall cabinets installed under a previously built drop ceiling area are square in design. Installed side by side as a unit their top line is straight. Unfortunately, the framing used to create the drop ceiling may be less than straight, and so an irregularity may exist. If this is true, a cove or shoe molding can be used to cover the joint and minimize the irregularity.

First of all, if the back wall has irregularities, such as depressions

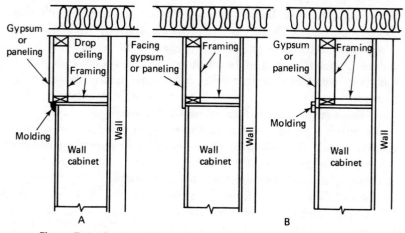

Figure 7–5 Blending Cabinet Units to Drop Ceilings—Three Methods

and protrusions, the straightedge can help in locating these areas. Secondly, the wall's surface must be examined for plumb. The fractional "out of plumb" area must be found and corrected. What steps should be taken if parts of the wall are plumb and others are not and there are either protrusions or depressions? Answer: shims cut to exact proportions should be installed in depressions since it is highly unlikely that the protrusions can be eliminated. But shims can also be installed for out-of-plumb spots. Most cabinets are fabricated with backs so the shims will not show. With shims in place the cabinets can be installed. Then each cabinet will hang true along the facing line, set plumb, and be adequately secured.

Let us recall the major principles of cabinet installation developed thus far. Most of the work requires that a proper foundation be established for each cabinet. This consists of straight, level and plumb conditions for floor and wall. Secondly, anchoring material and methods must be selected and provided. Third, careful and exact craftsmanship must maintain line and design features during the installation. Fourth, blending of the unit to its surroundings must be well thought out, planned, and executed. We shall see that these principles are fundamental to all types of cabinet work as we continue with the commercial display cabinet.

Display Cabinet

Display cabinets usually have glass fronts and tops and may have glass or plywood panel doors and solid ends. They may be fabricated with a solid base or be set on legs. We are concerned with those having solid bases. These bases are usually separate from the major case section. Sometimes one base provides the foundation for several sections. A solid base must be anchored to the floor which may be wood over wood joist. In that case anchors are simply screwed to the floor from inside of base to secure it to the floor. If, however, the base is installed over a concrete floor more work is involved.

In either type of installation the base must be absolutely level when installed. We examined the principles for surveying the floor and making judgements on its condition earlier. Let's use an example for review purposes. Assume that a ten-foot length of display cabinets is to be installed. Assume further that only one base is to be installed and on a concrete floor and that the cases will all be set on top of this base. Close examination of the floor where the base is to be installed reveals no significant defects. So we set the base in place and check it for level through its length and across both ends. It proves to be out of level by ¼″ through its length and across one end. Accordingly, we level the base with the aid of a shim and scribe along the floor. Using a power saw with a fine, sharp blade the stock is cut away.

Again we position the base and check for level. Asssume that it's level. How shall we anchor it? Any of three methods come to mind; they include the following:

1. Use a stud driver and anchor angle irons to the floor.
2. Mark the position of each angle iron on the floor; use an electric drill and cement-type bit to drill holes for the anchors; fasten.
3. Cut two 2×4 studs, one for each side, and use a stud driver to anchor them to the floor; then screw the base to the 2×4's.

Methods 1, 2, or 3 could be used successfully and the cases can be set in place after the base is installed. Any one would do a satisfactory job. Of the three, Method 2 requires more time.

There are numerous style variations and arrangements that may require special preparations. To name a few: a) the joining of cabinets where corners are turned, 30 degrees, 45 degrees, 60 degrees, or 90 degrees; b) outside corners to be installed; c) fillers to be installed; d) electrical fixtures to be reckoned with. These requirements really demand the best work from a craftsman because installation work can not show. Trimming and fitting must not mar any previously finished work; both the glue and screws must be carefully hidden. Cabinet design lines must be accurately maintained at the same time. These tasks make up the balance of the work of installing display cases.

Display Shelves and Hanging Racks

We will study the sample unit shown in Figure 7–6. Note that it consists of two distinct types: the unit at the left has a base storage cabinet with adjustable shelves above; the unit at the right has a low base and a rod across the upper portion on which clothing is hung. They both have a cap section behind which lighting is installed.

The standard practices developed earlier in this chapter are used for installation of display units. With preparations completed, the base unit is set in place after careful inspection. It is fastened to the wall and floor so that no fasteners show. The back upper portion of the unit is secured to framing stock or block walls. If the back consists of a peg board the material must be installed with space between the peg board and wall. For adjustable shelves, brackets are installed over the unit back. For clothes rods, standouts are installed as required near the top of the unit's back.

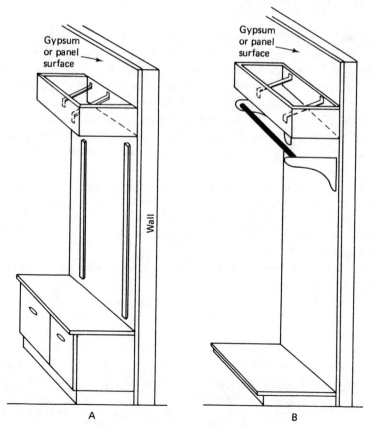

A   B

Figure 7–6 Typical Display Shelves and Racks

The head section with or without lighting may be prefabricated along with the rest of the unit, then the carpenter installs it. (If it must be fabricated on the job, see page 232 for data on custom cabinet-making.) Frequently, metal angle and strap braces are used to support the front of the head section. These are easily cut and installed. The back of the head section (if there is one) is fastened to the wall. The subcontracting cabinetmaker will, in most instances, provide installation suggestions.

Obviously, in this text we have not examined all the styles, types and variations of cabinets that can be installed. We have, however, defined the principles that are fundamental to every installation. To help in reinforcing your understanding of these principles we shall examine their application to cabinet construction.

CABINET CONSTRUCTION PRINCIPLES AFFECTING INSTALLATION

First, we recognize that much fine joinery and special gluing go into the construction of every cabinet. Variously, joints are butted, mitered, dadoed, dowelled, and mortised and tenoned. After this painstaking preparation glue is applied either to sections or to the whole cabinet. Then the work is clamped carefully to secure it and hold it square until the glue dries. This is most important if *no* racking or twisting of any part of a cabinet is to take place. Moreover, if later an installer applies a wrong force during his work, any one or several of the joints are sure to split.

If the wall or other area where the installation is to be made is out of level or not plumb and if the cabinet is made to conform to the irregularity, not only will its joints split but its appearance will suffer. This construction characteristic emphasizes the importance of shimming and trimming during installation.

Cabinet doors and drawers are made to very close tolerances. This means that the space between door edge and facing must be and is very small. So the least amount of racking by an installer produces misfits, such as a door that fails to close, or a door that hits or touches part of the facing and appears warped, or a drawer that will not close or whose front appears warped.

Doors and drawers must be fabricated and fitted square to their openings and installed accordingly. Generally today a cabinet door is made from a single piece of wood or plywood so it can only be made off the square by wrong cutting. Drawers which consist of a front, sides, back, and bottom each are constructed with either dadoed or dovetailed joinery. Either type tends to hold square; but like the solid door the drawer may be forced out of square if the joints are split. This, of course, destroys the drawer.

Once again we see that the cabinet should not be forced in any position that can make the door(s) and/or drawers fit improperly. These two parts of a cabinet usually can not be altered successfully except by re-cutting.

Remember that fastening a wall cabinet or unit in place was studied earlier in this chapter in considerable detail. It should be noted too that the cabinet's design features help keep it on the wall even though the considerable forces of dead weight and pull are always present. Dadoed shelves provide the strongest kind of resistance to weight and thrust because the shelf is made a part of the sides. See detail in the upper left of Figure 7–7. Usually strips of stock are incorporated at the bottom back and top back as built-in ledgers for anchoring purposes. See detail in upper right of Figure 7–7; the small arrows indicate the glued surfaces. This joining of surfaces by glue creates a sound anchoring surface for the cabinet. The back is rabbeted into the sides and the facings are dowelled (lower details in Figure 7–7).

If there are no such anchoring strips in the cabinets, the ledger board installation must be used. Again, dadoed shelves and sides combine to develop the total strength. Screws installed through the shelves into the ledgers hold the cabinet in place; the lower ledger supports the downward force, the top ledger takes the pull force.

Since many contemporary cabinets are made from particleboard, wood chips, or wood dust and glue, it is important to realize that these materials do not have the same strength as wood stock. For example, edge nailing into these cores usually results in splitting. Screws driven into their edge do not have great holding power either. Any work done near the edge of a piece of particleboard is in danger of splitting; therefore, great care must be used during installation. If screws are to be installed through shelves or sides, it is absolutely essential that pre-drilling starts as far back from the edge as possible. Such drilling minimizes the chances for splitting. Keeping screw heads away from the edge minimizes the chance of tearing surfaces as well.

The data just presented reemphasizes the installation techniques discussed earlier. Understanding construction principles will also aid in understanding custom cabinetmaking.

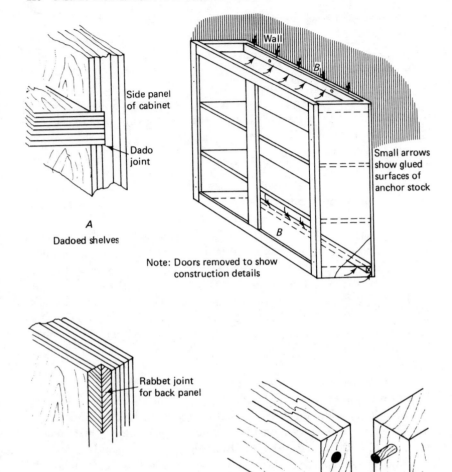

Side panel of cabinet

Dado joint

Wall

B

Small arrows show glued surfaces of anchor stock

A
Dadoed shelves

B

Note: Doors removed to show construction details

Rabbet joint for back panel

Rail

Dowelled cabinet frame

Stile

Figure 7–7 Layout for Wall Cabinet (Force and Design Details)

CUSTOM CABINETMAKING

Within the scope of work treated in this book the practicality of custom building cabinets narrows down to three types—counter-divider, wall storage, and window display.

Counter-Divider Cabinet

The counter-divider type of cabinet can be made from 28″ to 42″ high. A base unit, it has a counter top and may have shelves, drawers, doors, partitions, and other design features. Figure 7–8 shows an example of this type of cabinet; also included are cross-section views of the end and exploded views of selected joints.

Cabinets of this type are subject to lateral forces, such as persons leaning against them and bumping into them. Sound anchoring is, therefore, essential. In addition, proper cabinet joinery aids stability. The base is made from 2-inch nominal stock to desired width, length, and height. Temporary braces installed across the top keep the base section perfectly square until the base is anchored to the floor by sound methods.

To lay out the vertical pieces of the front, ends, and intermediate partitions the cabinetmaker again uses the plans. He installs a *planer* saw blade into a power saw and cuts the panels. Then he lays out and makes the rabbet and dado cuts as needed. In this work a router with dado bit is extremely useful. It saves time, does an accurate job, and actually results in cost savings.

The preliminary fitting is made when the vertical pieces are cut out. Then, from these pieces and the plans the intermediate horizontal pieces are laid out and cut. When the cuts are all made, the various sections are assembled without glue. If all the pieces fit properly, they are disassembled, glue is applied and they are reassembled.

The fastening is done with nails; 3d, 4d, 6d finish can be used. They must be driven in so that nail heads are concealed. The toe-nailing technique must be used in the following places:

* Underneath shelves
* Between the edge of a dado and panel exterior (toenailed into the shelf or vertical piece)

Remember that the nails provide holding power until the glue dries, but the glue is really the primary holding agent.

The next step in the construction is the preparation of the cabinet face. For counters this is usually the back of the cabinet; that is, where the employees operate. Again using the plans, the cabinetmaker cuts and dresses the stiles and rails. All these face pieces are joined with butt joints and dowels. A dowel jig makes the work very easy. Pieces of facing are aligned as required, marked for cutting, cut, and drilled for dowels. Then the facing is assembled dry. If everything fits well, it is positioned against the previously installed

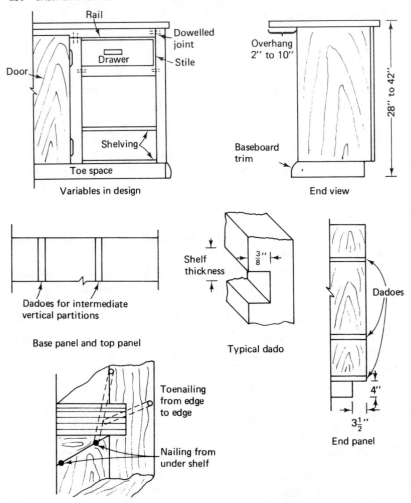

Figure 7–8 Layout for Typical Counter Cabinet

cabinet panels; if that all fits, it is disassembled, glued, reassembled, positioned against cabinet panels, and secured with glue clamps and toenails. Sure that everything is square, the cabinetmaker then braces the work as required.

He follows this work by installing the drawer guides. Then he prepares the top and installs it by whatever method is appropriate. No fasteners must show. However, if a plastic top is to be glued onto a plywood or particleboard underlayment, the underlayment can be topnailed with finish nails.

The final construction tasks that remain are application of trim, the fabricating, fitting, and installing of drawers, and the cutting, fitting, and installing of doors. These tasks require special skills because of the required joinery. Trim may range from simple to ornate moldings. They are mitered and glued in place. Doors may require rabbeting, close fitting (for flush doors), mortising for hinges in the door and facing, and the installing of cabinet locks. Drawer fronts are first prepared like a door. Then additional rabbeting and dadoing are used to make sides and bottom join to the front properly. At the rear of the side pieces a dado is made so that the back piece will fit into it. Grooves are also made near the bottom of the side pieces for the drawer bottom slides. Finally, the back piece is cut to fit and the drawer is assembled.

We can see from this description that there are many individual, highly technical, and somewhat difficult tasks to undertake during the custom building of a counter type cabinet. This review only identified the most significant joinery skills. Special uses are made of the saw, and a router is highly desirable for making dadoes and rabbets. The wall unit in the next study requires application of many of the same skills just examined.

Full-Length Wall Unit with Adjustable Shelving

The typical wall unit for this study has the characteristics seen in Figure 7–9: a standard base, a 14" high storage area with door, adjustable shelves to 6-ft high, and a cabinet crown to enclose the lighting.

The base is prepared as before. But, since it is placed against the wall, it is possible to nail it to wall framing. The front of the base should be anchored to the floor. Ensure that the base is level and square during its fabrication and installation. Trim stock as required to make the installation accurate.

The storage area should be built next. The construction principles for building a counter cabinet apply for building the storage cabinet section. Without going into all the detail again, the assembly consists of laying out and cutting the storage cabinet's bottom shelf, upright ends and partitions, and top. These pieces are glued and held in place with toenailing. Next the facings are cut, planed, joined with dowels and glue, and fastened to the cabinet. Finally, the doors are cut and installed.

The shelving area is built next. If the back or wall surface is not finished that should be done before the shelf brackets are installed.

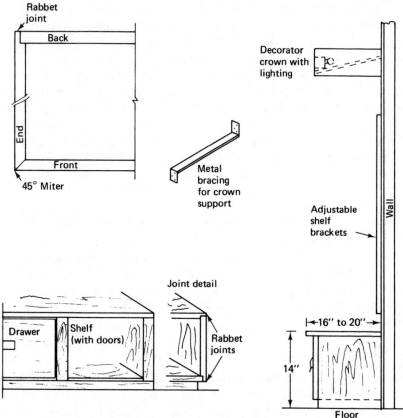

Figure 7–9 Layout for Typical Wall Display Unit

The brackets are screwed to studs on a wood frame wall. On a concrete block wall, however, plugs are installed or furring is fastened behind the cabinet back, then the shelf brackets are installed. The spacing of the shelf brackets should be based upon the weight each shelf is expected to carry; the more weight, the closer the shelf brackets.

The cabinet crown is added last. It may be a simple 1×8 as shown in Figure 7–9. It is held in place with angle braces screwed to the wall. On the back side of the crown fluorescent fixtures are installed. Joints in the crown must be made with the 45 degree miter. The outside piece of crown is held in place with bracing to the wall.

In this type of cabinetmaking, the hardware and some pre-finished materials are included in the total fabrication. In addition, any special bracing features must be designed. The principles of cabinet joinery are used throughout its construction.

Display Window

Most display windows are large glazed expanses with nothing more than a narrow platform behind (as seen from the street). Smaller shops and some restaurants use the area differently, however, since these establishments often sell special small-sized merchandise or display goods in a setting intended to create special effects. In such cases, the cabinet maker may play a significant part in the window design. His work could consist of making simple to complex shelving that incorporates wood, plastic, metal, and other materials. In the main though, his cabinetmaking techniques are simply employed for dados and miters.

Figure 7–10 shows a set of window display shelves. Stock for the vertical and horizontal pieces should be pre-cut to the required width. Layouts for the uprights or ends should be made first to full height and with the shelf dadoes all located. Next, the shelves themselves should be laid out to length. If dadoes are needed for intermediate partitions they should also be laid out. It is very important that all measurements be made with the same ruler and from the same reference point.

Cutting and dadoing follow the layout; trial assembly is next; and when all pieces fit as required, the final assembly should be done.

Figure 7–10 Shelving in Window Display Area

It may be that the shelf edges are unfinished, if so trim may have to be added. Usually, if the set of shelves is installed between framing members a casing is applied over the joint between cabinet member and wall covering. The side facing the window may also require a molding to cover this space. This molding may either be installed around the perimeter of the glass before the cabinet is slid in place or it may be attached to sides, top, and bottom if it is assembled within the window opening. Note that glass is usually installed after cabinet or shelves are in place and stained and varnished.

Conclusions

The basics of cabinetmaking are adaptable to every type of cabinet that is needed. We have examined three different areas on a commercial job that could employ on-site craftsmen to build cabinets. A high degree of skill and cabinetmaking knowledge is required to produce an excellent product. Layout must include understanding of wood, its strength and weakness, types of joints that produce the best strength and greatest reliability as well as look good and be an economical use of materials. Cutting and fitting are very exacting tasks. For instance, the under cut used frequently in trimming operations is *not* acceptable in cabinetmaking. Face-nailing is prohibited, but toenailing is acceptable if the heads do not show or can be covered. Glue is the primary holding agent. Finally, sanding must be mentioned. It is required to complete cabinetmaking in preparation for staining or painting.

CABINET WORK TRAINING PHASE

On-the-job cabinetmaking training during commercial construction is very limited. Few opportunities exist, so little if any cabinetmaking is done on-site. This is partly because subcontractors usually supply and install the cabinets and partly because cabinetmaking is a highly skilled and specialized part of woodworking. However, occasions for installing cabinets or units of cabinets are common. Therefore, training and then practical experience can be given to those in need.

To determine the training that each person needs, the cabinet work to be done must be identified. From this and the data already provided in this chapter, each carpenter likely to do cabinet work can be evaluated. Both knowledge and skill are entered on the training records. Figure 7–11 provides sample entries which include

Name of employee	Skill	Installation preparation	Base cabinets	Wall cabinets	Display racks and shelves	Trimming cabinets	Custom cabinets base units	Custom cabinets wall units	Custom cabinets display windows	Custom cabinet display shelving	Cabinet joinery principles	Remarks

□ None   ◩ Some   S Specialist

◪ Basic   ⊠ Skilled   A Apprentice

Figure 7–11 Training Record for Cabinet Work Crew

layout, inspection, prefabrication of pieces, assemblies, and various installations.

Training on principles of cabinet construction should be given as home study, and specifically for the types of cabinets or cabinet units that are to be built and/or installed on the current job. To aid in this learning effort special task activities follow.

CAB 1 TASK ACTIVITY: INSTALLATION TECHNIQUES

RESOURCES

Estimated Manhours for 1 man: townhouse kitchen or bathroom vanity
1 hr per lin ft of wall
commercial base unit
1 hr per 4 ft of cabinet
commercial wall unit
1–4 hrs per lin ft of wall

Materials:

prefabricated cabinets	shimming stock
anchors	ledgers

Tools:

level	ruler
straightedge	hammer
saw	plane
drill and bits	set of wood chisels
screwdriver	stud driver (*optional*)

PROCEDURES

1. With floor and elevation plans locate the exact position where the cabinets are to be installed; define with chalk or pencil marks.

2. Check surface areas with straightedge and level; record all surface irregularities. Select best repair method to eliminate each irregularity.

3. Prepare cabinets for installation in sequence:
   a. Pre-position cabinet; try out for fit
   b. Mark for anchoring
   c. Install shims and/or trim cabinet
   d. Install cabinet
   e. Join cabinets, ensuring that facings align and are all level
4. When required prepare base for counter-top anchoring; install top.
5. Add trim as required to complete installation.

CAB 2 TASK ACTIVITY: CABINETMAKING LAYOUT PRINCIPLES

RESOURCES

Estimated Manhours for 1 man: 8–40 hrs (avg.) per cabinet

Materials:

plans and specifications   plywoods
paneling   stock lumber
hardware

Tools:

framing square   straightedge   ruler

PROCEDURES

1. Study the plans and specifications to define structural elements of cabinet. List pieces by size and name.
2. Combine all pieces of the same material on a standard size panel of plywood to determine the most economical method of layout; be sure wood grain on each piece runs proper way.
3. Repeat Step 2 for stock lumber pieces, paneling, and backing material as required.
4. After cutting out all needed pieces, make layouts for dado, rabbet joints, and miter cuts.
5. Cut all required joinery and fit dry (no glue). Proceed to the next task activity.

CAB 3 TASK ACTIVITY: ASSEMBLING CABINETS BUILT ON-SITE

RESOURCES

Estimated Manhours for 1 man: 6 hrs per 48 cu ft of cabinet

Materials:

drawings and plans
facings and trim
sand paper
cabinet pieces
glue
nails, screws

Tools:

hammer	handsaw
level	set of wood chisels
screwdrivers	framing square
ruler	combination square

PROCEDURES

1. Position the base pieces to determine the accuracy of level; adjust for level by using shims. Scribe to the floor; cut along the scribed line. Assemble the pieces and install.
2. Lay cabinet bottom over base, position accurately, secure. Install end pieces, intermediate partitions, and intermediate shelves; hold together with cleat across top of cabinet.
3. Prepare facing piece. Assemble pieces dry and overlay against cabinet. Make any required corrections. Assemble pieces with glue in all joints and glue to cabinet. Toenail with 3d finish nails to hold facings until glue dries.
4. Install all drawer runners.
5. Build all drawers and doors. Fit and hang doors and attach hardware.
6. Install counter or cabinet top.

REVIEW QUESTIONS

1. Why should pre-built cabinets be treated as a long-lead item?
2. What is the usual specification for toe space?
3. What should you examine when surveying for a base cabinet installation?
4. When should scribing be used during cabinet installation?
5. Why must irregularities be eliminated in cabinets before the counter top is installed?
6. How is a typical counter top secured to a base cabinet?
7. What are the two main forces that must be taken into account when installing a wall cabinet?
8. How does sounding a wall aid in wall cabinet installation?
9. What is the advantage of using a lower ledger strip to support wall cabinets?
10. Describe the three methods of blending wall cabinets to drop or false ceilings.
11. Does installing base cabinets and display shelves include the use of base and wall cabinet installation techniques? Explain.
12. What makes a dadoed shelf so strong?
13. What are rails and styles and where are they used?
14. Where can nails be used in cabinetmaking?
15. Why should drawer guides be installed before counter tops are installed?

APPENDIX A

Working Stresses
For Joists and Rafters

(COURTESY OF NATIONAL FOREST PRODUCTS ASSOCIATION)

*Full data on span tables for joists and rafters, along with the 1981 supplement, are available for sale by the NFPA, 1250 Connecticut Avenue NW, Suite 200, Washington, DC 20036

Table W-1 Design Values For Joists and Rafters — Visual Grading

These "Fb" values are for use where repetitive members are spaced not more than 24 inches. For wider spacing, the "Fb" values should be reduced 13 percent.

Values for surfaced dry or surfaced green lumber apply at 19 percent maximum moisture content in use.

Species and grade	Size	Normal duration	Design value in bending "Fb" Snow loading	7-day loading	Modulus of elasti-city "E"	Grading Rules
Douglas Fir—Larch (Surfaced dry or surfaced green)						
Dense select structural		2800	3220	3500	1,900,000	
Select structural		2400	2760	3000	1,800,000	
Dense No. 1		2400	2760	3000	1,900,000	
No. 1 & appearance		2050	2360	2560	1,800,000	
Dense No. 2	2x4	1950	2240	2440	1,700,000	Western
No. 2		1650	1900	2060	1,700,000	Wood
No. 3		925	1060	1160	1,500,000	Products
Stud		925	1060	1160	1,500,000	Assn.
						(See
Construction		1200	1380	1500	1,500,000	notes 1
Standard	2x4	675	780	840	1,500,000	and 3)
Utility		325	370	410	1,500,000	
Dense select structural		2400	2760	3000	1,900,000	West
Select structural		2050	2360	2560	1,800,000	Coast
Dense No. 1	2x5	2050	2360	2560	1,900,000	Lumber
No. 1 & appearance	and	1750	2010	2190	1,800,000	Insp.
Dense No. 2	wider	1700	1960	2120	1,700,000	Bureau
No. 2		1450	1670	1810	1,700,000	
No. 3		850	980	1060	1,500,000	
Stud		850	980	1060	1,500,000	
Douglas Fir—Larch (North) (Surfaced dry or surfaced green)						
Select structural		2400	2760	3000	1,800,000	
No. 1 & appearance		2050	2360	2560	1,800,000	
No. 2	2x4	1650	1900	2060	1,700,000	Nat'l
No. 3		925	1060	1160	1,500,000	Lumber
Stud		925	1060	1160	1,500,000	Grades
						Auth.
Construction		1200	1380	1500	1,500,000	(A Canadian
Standard	2x4	675	780	840	1,500,000	Agency-
Utility		325	370	410	1,500,000	See
						notes 1,
Select structural		2050	2360	2560	1,800,000	2 and 3)
No. 1 & appearance	2x5	1750	2010	2190	1,800,000	
No. 2	and	1450	1670	1810	1,700,000	
No. 3	wider	850	980	1060	1,500,000	
Stud		850	980	1060	1,500,000	

Table W-1 Design Values For Joists and Rafters — Visual Grading (Cont)

These "Fb" values are for use where repetitive members are spaced not more than 24 inches. For wider spacing, the "Fb" values should be reduced 13 percent.

Values for surfaced dry or surfaced green lumber apply at 19 percent maximum moisture content in use.

Species and grade	Size	Normal duration	Design value in bending "Fb" Snow loading	7-day loading	Modulus of elasticity "E"	Grading Rules
Douglas Fir South (Surfaced dry or surfaced green)						
Select structural		2300	2640	2880	1,400,000	
No. 1 & appearance		1950	2240	2440	1,400,000	Western
No. 2		1600	1840	2000	1,300,000	Wood
No. 3	2x4	875	1010	1090	1,100,000	Products
Stud		875	1010	1090	1,100,000	Assn.
Construction		1150	1320	1440	1,100,000	(See
Standard	2x4	650	750	810	1,100,000	notes 1
Utility		300	340	380	1,100,000	and 3)
Select structural		1950	2240	2440	1,400,000	
No. 1 & appearance	2x5	1650	1900	2060	1,400,000	
No. 2	and	1350	1550	1690	1,300,000	
No. 3	wider	800	920	1000	1,100,000	
Stud		800	920	1000	1,100,000	

Table W-1 Design Values For Joists and Rafters — Visual Grading (Cont)

These "Fb" values are for use where repetitive members are spaced not more than 24 inches. For wider spacing, the "Fb" values should be reduced 13 percent.

Values for surfaced dry or surfaced green lumber apply at 19 percent maximum moisture content in use.

Species and grade	Size	Normal duration	Design value in bending "Fb" Snow loading	7-day loading	Modulus of elasti- city "E"	Grading Rules
Southern Pine (Surfaced dry)						
Select structural		2300	2640	2880	1,700,000	
Dense select structural		2700	3100	3380	1,800,000	
No. 1		1950	2240	2440	1,700,000	
No. 1 dense		2300	2640	2880	1,800,000	
No. 2	2x4	1650	1900	2060	1,600,000	
No. 2 dense		1900	2180	2380	1,600,000	
No. 3		900	1040	1120	1,400,000	
No. 3 dense		1050	1210	1310	1,500,000	Southern
Stud		900	1040	1120	1,400,000	Pine
						Insp.
Construction		1150	1320	1440	1,400,000	Bureau
Standard	2x4	675	780	840	1,400,000	
Utility		300	340	380	1,400,000	(See
						note 3)
Select structural		2000	2300	2500	1,700,000	
Dense select structural		2350	2700	2940	1,800,000	
No. 1		1700	1960	2120	1,700,000	
No. 1 dense	2x5	2000	2300	2500	1,800,000	
No. 2	and	1400	1610	1750	1,600,000	
No. 2 dense	wider	1650	1900	2060	1,600,000	
No. 3		800	920	1000	1,400,000	
No. 3 dense		925	1060	1160	1,500,000	
Stud		850	980	1060	1,400,000	

Table W-1 Design Values For Joists and Rafters — Visual Grading (Cont)

These "Fb" values are for use where repetitive members are spaced not more than 24 inches. For wider spacing, the "Fb" values should be reduced 13 percent.

Values for surfaced dry or surfaced green lumber apply at 19 percent maximum moisture content in use.

Species and grade	Size	Normal duration	Design value in bending "Fb" Snow loading	7-day loading	Modulus of elasti- city "E"	Grading Rules
Southern Pine (Surfaced at 15 percent maximum moisture content-KD)						
Select structural		2500	2880	3120	1,800,000	
Dense select structural		2900	3340	3620	1,900,000	
No. 1		2100	2420	2620	1,800,000	
No. 1 dense		2450	2820	3060	1,900,000	
No. 2	2x4	1750	2010	2190	1,600,000	
No. 2 dense		2050	2360	2560	1,700,000	Southern
No. 3		975	1120	1220	1,500,000	Pine
No. 3 dense		1150	1320	1440	1,500,000	Insp.
Stud		975	1120	1220	1,500,000	Bureau
Construction		1250	1440	1560	1,500,000	
Standard	2x4	725	830	910	1,500,000	
Utility		300	340	380	1,500,000	(See note 3)
Select structural		2150	2470	2690	1,800,000	
Dense select structural		2500	2880	3120	1,900,000	
No. 1		1850	2130	2310	1,800,000	
No. 1 dense	2x5	2150	2470	2690	1,900,000	
No. 2	and	1500	1720	1880	1,600,000	
No. 2 dense	wider	1750	2010	2190	1,700,000	
No. 3		875	1010	1090	1,500,000	
No. 3 dense		1000	1150	1250	1,500,000	
Stud		900	1040	1120	1,500,000	

Table W-1 Notes Applicable to Visually Graded Joists and Rafters

1. When 2″ lumber is manufactured at a maximum moisture content of 15% (grade-marked MC-15) and used in a condition where the moisture content does not exceed 15% the design values shown in Table W-1 for "surfaced dry or surface green" lumber may be increased eight percent (8%) for design value in bending "Fb", and five percent (5%) for Modulus of Elasticity "E".

2. National Lumber Grades Authority is the Canadian rules writing agency responsible for preparation, maintenance and dissemination of a uniform softwood lumber grading rule for all Canadian species.

3. Design values for Stud grade in 2 x 5 and wider size classifications apply to 5″ and 6″ widths only.

APPENDIX   B

Maximum   Spans
For   Joists   and   Rafters

*(SELECTED TABLES AND DATA FROM TECHNICAL BULLETIN NO. 2
BY COURTESY OF SOUTHERN FOREST PRODUCTS ASSOCIATION)*

TABLE NO. 1. FLOOR JOISTS—30 psf live load. Sleeping rooms and attic floors. (Spans shown in unshaded columns are based on a deflection limitation of $l/360$. Spans shown in shaded columns are limited by the recommended extreme fiber stress in bending value of the grade and includes a 10 psf dead load.)

Size and Spacing In.	Grade[2] In. o.c.	Dense Sel Str KD and No.1 Dense KD	Dense Sel Str KD, Sel Str No.1 Dense and No.1 KD	Sel Str, No.1 and No.2 Dense KD	No.2 Dense, No.2 KD and No.2	No.3[3] Dense KD	No.3[3] Dense	No.3 KD	No.3
2 x 6	12.0	12-6	12-3	12-0	11-10	11-3	10-11	10-5	10-1
	13.7	11-11	11-9	11-6	11-3	10-6	10-3	9-9	9-5
	16.0	11-4	11-2	10-11	10-9	9-9	9-6	9-0	8-9
	19.2	10-8	10-6	10-4	10-1	8-11	8-8	8-3	8-0
	24.0	9-11	9-9	9-7	9-4	8-0	7-9	7-4	7-1
2 x 8	12.0	16-6	16-2	15-10	15-7	14-10	14-5	13-9	13-3
	13.7	15-9	15-6	15-2	14-11	13-11	13-6	12-10	12-5
	16.0	15-0	14-8	14-5	14-2	12-10	12-6	11-11	11-6
	19.2	14-1	13-10	13-7	13-4	11-9	11-5	10-10	10-6
	24.0	13-1	12-10	12-7	12-4	10-6	10-2	9-9	9-5
2 x 10	12.0	21-0	20-8	20-3	19-10	18-11	18-5	17-6	16-11
	13.7	20-1	19-9	19-4	19-0	17-9	17-2	16-5	15-10
	16.0	19-1	18-9	18-5	18-0	16-5	15-11	15-2	14-8
	19.2	18-0	17-8	17-4	17-0	15-0	14-6	13-10	13-5
	24.0	16-8	16-5	16-1	15-9[1]	13-5	13-0	12-5	12-0
2 x 12	12.0	25-7	25-1	24-8	24-2	23-0	22-4	21-4	20-7
	13.7	24-5	24-0	23-7	23-1	21-7	20-11	19-11	19-3
	16.0	23-3	22-10	22-5	21-11	19-11	19-4	18-6	17-10
	19.2	21-10	21-6	21-1	20-8	18-3	17-8	16-10	16-3
	24.0	20-3	19-11	19-7	19-2[1]	16-3	15-10	15-1	14-7

1. The span for No 2 grade 24 inches o.c. spacing is 2x10 15-8; 2x12 19-1.
2. Terms and abbreviations Sel Str means select structural KD means KD15 dried to a moisture content of 15 or less where KD is not shown the material is dried to a moisture content of 19 or less. Lumber dried to 19 or less will be stamped S-DRY or KD S-DRY
3. These grades may not be commonly available

TABLE NO. 2. FLOOR JOISTS—40 psf live load. All rooms except sleeping rooms and attic floors. (Spans shown in unshaded columns are based on a deflection limitation of $l/360$. Spans shown in shaded columns are limited by the recommended extreme fiber stress in bending value of the grade and includes a 10 psf dead load.)

Size and Spacing in in. o.c. (Grade[2])	Dense Sel Str and No. 1 KD	Dense Sel Str, Sel Str KD, No. 1, Dense and No. 1 KD	Sel Str, No. 1 and No. 2 Dense KD	No. 2 Dense, No. 2 KD and No. 2	No. 3[3] Dense KD	No. 3[3] Dense	No. 3 KD	No. 3
2 x 6 12.0	11-4	11-2	10-11	10-9	10-1	9-9	9-4	9-0
13.7	10-10	10-8	10-6	10-3	9-5	9-2	8-9	8-5
16.0	10-4	10-2	9-11	9-9	8-9	8-6	8-1	7-10
19.2	9-8	9-6	9-4	9-2	8-0	7-9	7-4	7-1
24.0	9-0	8-10	8-8	8-6[1]	7-1	6-11	6-7	6-4
2 x 8 12.0	15-0	14-8	14-5	14-2	13-3	12-11	12-4	11-11
13.7	14-4	14-1	13-10	13-6	12-5	12-1	11-6	11-1
16.0	13-7	13-4	13-1	12-10	11-6	11-2	10-8	10-3
19.2	12-10	12-7	12-4	12-1	10-6	10-2	9-9	9-5
24.0	11-11	11-8	11-5	11-3[1]	9-5	9-1	8-8	8-5
2 x 10 12.0	19-1	18-9	18-5	18-0	16-11	16-5	15-8	15-2
13.7	18-3	17-11	17-7	17-3	15-10	15-5	14-8	14-2
16.0	17-4	17-0	16-9	16-5	14-8	14-3	13-7	13-1
19.2	16-4	16-0	15-9	15-5	13-5	13-0	12-5	12-0
24.0	15-2	14-11	14-7	14-4[1]	12-0	11-8	11-1	10-9
2 x 12 12.0	23-3	22-10	22-5	21-11	20-7	20-0	19-1	18-5
13.7	22-3	21-10	21-5	21-0	19-3	18-9	17-10	17-3
16.0	21-1	20-9	20-4	19-11	17-10	17-4	16-6	16-0
19.2	19-10	19-6	19-2	18-9	16-3	15-10	15-1	14-7
24.0	18-5	18-1	17-9	17-5[1]	14-7	14-2	13-6	13-0

1. The span for No. 2 grade 24 inches o.c. spacing is 2x6 **8-4** 2x8 **11-0** 2x10 **14-0** 2x12 **17-1**.
2. Terms and abbreviations Sel Str means select structural. KD means KD15 dried to a moisture content of 15 or less where KD is not shown the material is dried to a moisture content of 19% or less. Lumber dried to 19% or less will be stamped S-DRY or KD S-DRY
3. These grades may not be commonly available

TABLE NO. 3. FLOOR JOISTS, CONCRETE SUBFLOOR—30 psf live load. Sleeping rooms. (Spans are based on a deflection limitation of l/360, except spans in shaded area are limited by recommended extreme fiber stress in bending value of the grade and includes a 27 psf dead load consisting of 17 psf for 2" of lightweight concrete and 10 psf for framing.)

Size and Spacing in.	Grade [2] in. o.c.	Dense Sel Str KD and No. 1 Dense KD	Dense Sel Str and Sel Str KD	Sel Str	No. 1 Dense	No. 1 KD	No. 2 Dense KD	No. 1	No. 2 Dense	No. 2 KD	No. 2	No. 3 Dense KD [3]	No. 3 Dense [3]	No. 3 KD	No. 3
2 x 6	12.0	12-6	12-3	12-0	12-3	12-3	12-0	12-0	11-10	11-6	11-1	9-5	9-2	8-9	8-5
	13.7	11-11	11-9	11-6	11-9	11-9	11-6	11-4	11-2	10-9	10-4	8-10	8-7	8-2	7-11
	16.0	11-4	11-2	10-11	11-2	11-1	10-10	10-6	10-4	10-0	9-7	8-2	7-11	7-7	7-4
	19.2	10-8	10-6	10-4	10-5	10-1	9-11	9-7	9-5	9-1	8-9	7-6	7-3	6-11	6-8
	24.0	9-11¹	9-9¹	9-5	9-4	9-0	8-11	8-7	8-5	8-2	7-10	6-8	6-6	6-2	6-0
2 x 8	12.0	16-6	16-2	15-10	16-2	16-2	15-10	15-10	15-7	15-2	14-7	12-5	12-1	11-6	11-1
	13.7	15-9	15-6	15-2	15-6	15-6	15-2	15-0	14-9	14-2	13-7	11-8	11-4	10-9	10-5
	16.0	15-0	14-8	14-5	14-8	14-7	14-4	13-10	13-7	13-2	12-7	10-9	10-5	10-0	9-8
	19.2	14-1	13-10	13-7	13-8	13-4	13-1	12-8	12-5	12-0	11-6	9-10	9-7	9-1	8-10
	24.0	13-1¹	12-10¹	12-5	12-3	11-11	11-8	11-4	11-1	10-9	10-4	8-10	8-6	8-2	7-10
2 x 10	12.0	21-0	20-8	20-3	20-8	20-8	20-3	20-3	19-10	19-4	18-7	15-10	15-5	14-8	14-2
	13.7	20-1	19-9	19-4	19-9	19-9	19-4	19-1	18-9	18-1	17-5	14-10	14-5	13-9	13-3
	16.0	19-1	18-9	18-5	18-9	18-7	18-3	17-8	17-5	16-9	16-1	13-9	13-4	12-9	12-3
	19.2	18-0	17-8	17-4	17-6	17-0	16-8	16-2	15-10	15-3	14-8	12-7	12-2	11-7	11-3
	24.0	16-8¹	16-5¹	15-10	15-8	15-2	14-11	14-5	14-2	13-8	13-2	11-3	10-11	10-5	10-0
2 x 12	12.0	25-7	25-1	24-8	25-1	25-1	24-8	24-8	24-2	23-6	22-7	19-4	18-9	17-10	17-3
	13.7	24-5	24-0	23-7	24-0	24-0	23-7	23-3	22-10	22-0	21-2	18-1	17-6	16-9	16-2
	16.0	23-3	22-10	22-5	22-10	22-7	22-3	21-6	21-2	20-4	19-7	16-9	16-3	15-6	14-11
	19.2	21-10	21-6	21-1	21-3	20-8	20-4	19-8	19-4	18-7	17-10	15-3	14-10	14-1	13-8
	24.0	20-3¹	19-11¹	19-4	19-0	18-5	18-2	17-7	17-3	16-8	16-0	13-8	13-3	12-8	12-2

1. The spans for No. 1 Dense KD and Select Structural KD. 24 inches o.c. spacing is 2x6, 9.8; 2x8, 12.9; 2x10, 16.4; 2x12, 19.10.
2. Terms and abbreviations: Sel. Str. means select structural. KD means structural KD15 dried to a moisture content of 15% or less where KD is not shown the material is dried to a moisture content of 19% or less. Lumber dried to 19% or less will be stamped S-DRY or KD S-DRY.
3. These grades may not be commonly available.

TABLE NO.4 CEILING JOISTS—PLASTER CEILING—10 psf live load. No future sleeping rooms and no attic storage, roof slopes 3 in 12 or less. (Spans are based on a deflection limitation of l/360, except spans shown in shaded areas are limited by the recommended extreme fiber stress in bending value of the grade and includes a 5 psf dead load.)

Size and Spacing in. in. o.c.	Grade	Dense Sel Str KD and No. 1 Dense KD	Dense Sel Str KD, No. 1 Dense and No. 1 KD	Sel Str, No. 1 and No. 2 Dense KD	No. 2 Dense, No. 2 KD and No. 2	No. 3 Dense KD[2]	No. 3 Dense	No. 3 KD	No. 3	Construction KD	Construction	Standard KD	Standard
2 x 4	12.0	11-6	11-3	11-1	10-10	10-7	10-7	10-7	10-4	10-7	10-4	9-11	9-6
	13.7	11-0	10-9	10-7	10-4	10-2	10-2	10-2	9-11	10-2	9-11	9-3	8-11
	16.0	10-5	10-3	10-0	9-10	9-8	9-8	9-8	9-5	9-8	9-5	8-7	8-3
	19.2	9-10	9-8	9-5	9-3	9-1	9-1	9-1	8-8	9-1	8-10	7-10	7-6
	24.0	9-1	8-11	8-9	8-7	8-5	8-5	8-2	7-9	8-5	8-3	7-0	6-9
2 x 6	12.0	18-0	17-8	17-4	17-0	16-8	16-8	16-8	16-3				
	13.7	17-3	16-11	16-7	16-3	15-11	15-11	15-11	15-5				
	16.0	16-4	16-1	15-9	15-6	15-2	15-2	14-9	14-3				
	19.2	15-5	15-2	14-10	14-7	14-3	14-1	13-6	13-0				
	24.0	14-4	14-1	13-9	13-6	13-0	12-8	12-0	11-8				
2 x 8	12.0	23-9	23-4	22-11	22-5	21-11	21-11	21-11	21-5				
	13.7	22-9	22-4	21-11	21-6	21-0	21-0	21-0	20-3				
	16.0	21-7	21-2	20-10	20-5	19-11	19-11	19-5	18-9				
	19.2	20-4	19-11	19-7	19-2	18-9	18-7	17-9	17-2				
	24.0	18-10	18-6	18-2	17-10	17-2	16-8	15-10	15-4				
2 x 10	12.0	30-4	29-9	29-2	28-7	28-0	28-0	28-0	27-5				
	13.7	29-0	28-6	27-11	27-5	26-10	26-10	26-9	25-11				
	16.0	27-6	27-1	26-6	26-0	25-5	25-5	24-10	23-11				
	19.2	25-11	25-5	25-0	24-6	23-11	23-9	22-8	21-10				
	24.0	24-1	23-8	23-2	22-9	21-10	21-3	20-3	19-7				

1. Terms and abbreviations. Sel Str means select structural. KD means KD15 dried to a moisture content of 15 or less, where KD is not shown the material is dried to a moisture content of 19% or less. Lumber dried to 19% or less will be stamped S-DRY or KD S-DRY

2. These grades may not be commonly available

TABLE NO. 5. CEILING JOISTS—Drywall ceiling—10 psf live load. No future sleeping rooms and no attic storage, roof slopes 3 in 12 or less. (Spans are based on a deflection limitation of l/240, except spans in shaded area are limited by the recommended extreme fiber stress in bending value of the grade and includes a 5 psf dead load.)

Size and Spacing in.	Grade[2] in. o.c.	Dense Sel Str KD and No. 1 Dense KD	Dense Sel Str, Sel Str No. 1 Dense and No. 1 KD	Sel Str, No. 1, and No. 2 Dense KD	No. 2 Dense, No. 2 KD and No. 2	No. 3 Dense KD	No. 3 Dense[3]	No. 3 KD	No. 3	Construction KD	Construction	Standard KD	Standard
2 x 4	12.0	13-2	12-11	12-8	12-5	12-2	12-0	11-6	11-0	12-2	11-10	9-11	9-6
	13.7	12-7	12-4	12-1	11-10	11-7	11-3	10-9	10-4	11-7	11-4	9-3	8-11
	16.0	11-11	11-9	11-6	11-3	10-10	10-5	10-0	9-6	11-0	10-9	8-7	8-3
	19.2	11-3	11-0	10-10	10-7	9-11	9-6	9-1	8-8	10-4	9-11	7-10	7-6
	24.0	10-5	10-3	10-0	9-10	8-10	8-6	8-2	7-9	9-3	8-10	7-0	6-9
2 x 6	12.0	20-8	20-3	19-11	19-6	18-5	17-10	17-0	16-5				
	13.7	19-9	19-5	19-0	18-8	17-2	16-8	15-11	15-5				
	16.0	18-9	18-5	18-1	17-8	15-11	15-6	14-9	14-3				
	19.2	17-8	17-4	17-0	16-8	14-6	14-1	13-6	13-0				
	24.0	16-4	16-1	15-9	15-6[1]	13-0	12-8	12-0	11-8				
2 x 8	12.0	27-2	26-9	26-2	25-8	24-3	23-6	22-5	21-8				
	13.7	26-0	25-7	25-1	24-7	22-8	22-0	21-0	20-3				
	16.0	24-8	24-3	23-10	23-4	21-0	20-5	19-5	18-9				
	19.2	23-3	22-10	22-5	21-11	19-2	18-7	17-9	17-2				
	24.0	21-7	21-2	20-10	20-5[1]	17-2	16-8	15-10	15-4				
2 x 10	12.0	34-8	34-1	33-5	32-9	30-11	30-0	28-8	27-8				
	13.7	33-2	32-7	32-0	31-4	28-11	28-1	26-9	25-11				
	16.0	31-6	31-0	30-5	29-9	26-9	26-0	24-10	23-11				
	19.2	29-8	29-2	28-7	28-0	24-5	23-9	22-8	21-10				
	24.0	27-6	27-1	26-6	26-0[1]	21-10	21-3	20-3	19-7				

1 The span for No. 2 grade, 24 inches o.c. spacing is 2x6 15-3, 2x8 20-1, 2x10 25-7.

2 Terms and abbreviations: Sel Str means select structural. KD means KD15, dried to a moisture content of 15 or less, when KD is not shown the material is dried to a moisture content of 19 or less. Lumber dried to 19 or less will be stamped S-DRY or KD S-DRY.

3 These grades may not be commonly available.

TABLE NO. 6. RAFTERS—Any slope—With drywall ceiling—30 psf live load. (Spans are based on a deflection limitation of I/240, except spans in shaded areas are limited by the recommended fiber stress in bending value of the grade and includes a 15 psf dead load.)

Size and Spacing in.	Grade[2] in. o.c.	Dense Sel Str KD	No. 1 Dense KD	Dense Sel Str and Sel Str KD	No. 1 Dense	Sel Str	No.1 KD	No. 2 Dense KD	No. 1	No. 2 Dense	No. 2 KD	No. 2	No.3 Dense KD[3]	No.3 Dense[3]	No.3 KD	No. 3
2 x 6	12.0	14-4	14-4	14-1	14-1	13-9	14-1	13-9	13-8	13-5	12-11	12-5	10-7	10-4	9-10	9-6
	13.7	13-8	13-8	13-5	13-5	13-2	13-5	13-2	12-9	12-7	12-1	11-8	9-11	9-8	9-2	8-11
	16.0	13-0	13-0	12-9	12-9	12-6	12-5	12-3	11-10	11-8	11-2	10-9	9-2	8-11	8-6	8-3
	19.2	12-3	12-2	12-0	11-8	11-9	11-4	11-2	10-10	10-7	10-3	9-10	8-5	8-2	7-9	7-6
	24.0	11-4	10-11	11-2[1]	10-6	10-7	10-2	10-0	9-8	9-6	9-2	8-10	7-6	7-3	6-11	6-9
2 x 8	12.0	18-10	18-10	18-6	18-6	18-2	18-6	18-2	18-0	17-8	17-1	16-5	14-0	13-7	12-11	12-6
	13.7	18-0	18-0	17-9	17-9	17-5	17-8	17-5	16-10	16-7	16-0	15-4	13-1	12-9	12-1	11-9
	16.0	17-2	17-2	16-10	16-10	16-6	16-5	16-2	15-7	15-4	14-9	14-2	12-1	11-9	11-3	10-10
	19.2	16-1	16-1	15-10	15-5	15-6	15-0	14-9	14-3	14-0	13-6	12-11	11-1	10-9	10-3	9-11
	24.0	15-0	14-5	14-8[1]	13-10	14-0	13-5	13-2	12-9	12-6	12-1	11-7	9-11	9-7	9-2	8-10
2 x 10	12.0	24-1	24-1	23-8	23-8	23-2	23-8	23-2	23-0	22-7	21-9	20-11	17-10	17-4	16-6	16-0
	13.7	23-0	23-0	22-7	22-7	22-2	22-7	22-2	21-6	21-2	20-4	19-7	16-8	16-3	15-6	14-11
	16.0	21-10	21-10	21-6	21-6	21-1	20-11	20-7	19-11	19-7	18-10	18-1	15-6	15-0	14-4	13-10
	19.2	20-7	20-6	20-2	19-8	19-10	19-1	18-9	18-2	17-10	17-2	16-6	14-1	13-8	13-1	12-8
	24.0	19-1	18-4	18-9[1]	17-7	17-10	17-1	16-10	16-3	16-0	15-5	14-9	12-8	12-3	11-8	11-4
2 x 12	12.0	29-3	29-3	28-9	28-9	28-2	28-9	28-2	27-11	27-6	26-6	25-5	21-9	21-1	20-1	19-5
	13.7	28-0	28-0	27-6	27-6	27-0	27-6	27-0	26-2	25-8	24-9	23-10	20-4	19-9	18-10	18-2
	16.0	26-7	26-7	26-1	26-1	25-7	25-5	25-0	24-3	23-9	22-11	22-0	18-10	18-3	17-5	16-10
	19.2	25-0	25-0	24-7	23-11	24-1	23-3	22-10	22-1	21-9	20-11	20-1	17-2	16-8	15-11	15-4
	24.0	23-3	22-4	22-10[1]	21-5	21-9	20-9	20-5	19-9	19-5	18-9	18-0	15-4	14-11	14-3	13-9

1. The span for Select Structural KD 24 inches o.c. spacing is : 2x6, 10-11; 2x8, 14-5; 2x10, 18-4; 2x12, 22-4.
2. Terms and abbreviations: Sel. Str. means select structural. KD means KD15, dried to a moisture content of 15%, where KD is not shown the material is dried to a moisture content of 19%, or less. Lumber dried to 19%, or less will be stamped S-DRY or KD S-DRY
3. These grades may not be commonly available

TABLE NO. 7. RAFTERS—Any slope—With drywall ceiling—40 psf live load. (Spans are based on a deflection limitation of l/240, except spans in shaded areas are limited by the recommended fiber stress in bending value of the grade and includes a 15 psf dead load.)

Size and Spacing in. / in. o.c.		Dense Sel Str KD	No. 1 Dense KD	Dense Sel Str and Sel Str KD	No. 1 Dense	Sel Str	No. 1 KD	No. 2 Dense KD	No. 1	No. 2 Dense	No. 2 KD	No. 2	No. 3 Dense KD[3]	No. 3 Dense	No. 3 KD	No. 3[3]
2 x 6	12.0	13-0	13-0	12-9	12-9	12-6	12-9	12-6	12-4	12-2	11-8	11-3	9-7	9-4	8-11	8-7
	13.7	12-5	12-5	12-3	12-3	12-0	12-2	12-0	11-7	11-4	10-11	10-6	9-0	8-9	8-4	8-0
	16.0	11-10	11-10	11-7	11-7	11-5	11-3	11-1	10-8	10-6	10-2	9-9	8-4	8-1	7-8	7-5
	19.2	11-1	11-0	10-11	10-7	10-8	10-3	10-1	9-9	9-7	9-3	8-11	7-7	7-4	7-0	6-9
	24.0	10-4	9-10	10-2¹	9-6	9-7	9-2	9-0	8-9	8-7	8-3	7-11	6-9	6-7	6-3	6-1
2 x 8	12.0	17-2	17-2	16-10	16-10	16-6	16-10	16-6	16-4	16-0	15-5	14-10	12-8	12-4	11-9	11-4
	13.7	16-5	16-5	16-1	16-1	15-9	16-0	15-9	15-3	15-0	14-5	13-10	11-10	11-6	11-0	10-7
	16.0	15-7	15-7	15-3	15-3	15-0	14-10	14-7	14-1	13-10	13-4	12-10	11-0	10-8	10-2	9-10
	19.2	14-8	14-7	14-5	13-11	14-1	13-6	13-4	12-11	12-8	12-2	11-9	10-0	9-9	9-3	8-11
	24.0	13-7	13-0	13-4¹	12-6	12-8	12-1	11-11	11-6	11-4	10-11	10-6	8-11	8-8	8-3	8-0
2 x 10	12.0	21-10	21-10	21-6	21-6	21-1	21-6	21-1	20-10	20-5	19-8	18-11	16-2	15-8	14-11	14-5
	13.7	20-11	20-11	20-6	20-6	20-2	20-5	20-1	19-5	19-1	18-5	17-8	15-1	14-8	14-0	13-6
	16.0	19-10	19-10	19-6	19-6	19-2	18-11	18-7	18-0	17-8	17-1	16-5	14-0	13-7	12-11	12-6
	19.2	18-8	18-7	18-4	17-10	18-0	17-3	17-0	16-5	16-2	15-7	14-11	12-9	12-5	11-10	11-5
	24.0	17-4	16-7	16-7¹	15-11	16-2	15-5	15-2	14-8	14-5	13-11	13-5	11-5	11-1	10-7	10-3
2 x 12	12.0	26-7	26-7	26-1	26-1	25-7	26-1	25-7	25-3	24-10	23-11	23-0	19-8	19-1	18-2	17-7
	13.7	25-5	25-5	25-0	25-0	24-6	24-10	24-6	23-8	23-3	22-5	21-6	18-5	17-10	17-0	16-5
	16.0	24-2	24-2	23-9	23-9	23-3	23-0	22-8	21-11	21-6	20-9	19-11	17-0	16-6	15-9	15-3
	19.2	22-9	22-7	22-4	21-8	21-11	21-0	20-8	20-0	19-8	18-11	18-2	15-6	15-1	14-5	13-11
	24.0	21-1	20-2	20-9¹	19-4	19-8	18-9	18-6	17-11	17-7	16-11	16-3	13-11	13-6	12-10	12-5

1. The span for Select Structural KD, 24 inches o.c. spacing is: 2x6, 9-10; 2x8, 13-0; 2x10, 16-2; 2x12, 20-2.
2. Terms and abbreviations: Sel Str means select structural. KD means KD15, dried to a moisture content of 15% or less. where KD is not shown the material is dried to a moisture content of 19% or less. Lumber dried to 19% or less will be stamped S-DRY or KD S-DRY.
3. These grades may not be commonly available.

TABLE NO. 8. RAFTERS—Low slope (3 in 12 or less)— With no finished ceiling — 20 psf live load. (Spans are based on a deflection limitation of l/240 except spans in shaded areas are limited by the recommended extreme fiber stress in bending value of the grade and includes a 10 psf dead load.)

Size and Spacing	Grade[3]	Dense Sel Str KD and No.1 Dense KD	Dense Sel Str, Sel Str KD, No.1, No.1 Dense and No.1 KD	Sel Str, No.2 Dense KD	No. 1	No. 2 Dense	No. 2 KD	No. 2	No. 3[4] Dense KD	No. 3[4] Dense	No. 3 KD	No. 3
in.	in. o.c.											
2 x 6	12.0	16-4	16-1	15-9	15-9	15-6	15-6	15-3	13-0	12-8	12-0	11-8
	13.7	15-8	15-5	15-1	15-1	14-9	14-9	14-3	12-2	11-10	11-3	10-11
	16.0	14-11	14-7	14-4	14-4	14-1	13-9	13-2	11-3	10-11	10-5	10-1
	19.2	14-0	13-9	13-6	13-3	13-0	12-6	12-0	10-3	10-0	9-6	9-2
	24.0	13-0	12-9[1]	12-6[2]	11-10	11-8	11-2	10-9	9-2	8-11	8-6	8-3
2 x 8	12.0	21-7	21-2	20-10	20-10	20-5	20-5	20-1	17-2	16-8	15-10	15-4
	13.7	20-8	20-3	19-11	19-11	19-6	19-6	18-9	16-0	15-7	14-10	14-4
	16.0	19-7	19-3	18-11	18-11	18-6	18-1	17-5	14-10	14-5	13-9	13-3
	19.2	18-5	18-2	17-9	17-5	17-2	16-6	15-10	13-7	13-2	12-7	12-1
	24.0	17-2	16-10[1]	16-6[2]	15-7	15-4	14-9	14-2	12-1	11-9	11-3	10-10
2 x 10	12.0	27-6	27-1	26-6	26-6	26-0	26-0	25-7	21-10	21-3	20-3	19-7
	13.7	26-4	25-10	25-5	25-5	24-11	24-11	24-0	20-6	19-10	18-11	18-4
	16.0	25-0	24-7	24-1	24-1	23-8	23-1	22-2	18-11	18-5	17-6	16-11
	19.2	23-7	23-2	22-8	22-3	21-10	21-1	20-3	17-3	16-9	16-0	15-6
	24.0	21-10	21-6[1]	21-1[2]	19-11	19-7	18-10	18-1	15-6	15-0	14-4	13-10
2 x 12	12.0	33-6	32-11	32-3	32-3	31-8	31-8	31-2	26-7	25-10	24-8	23-9
	13.7	32-0	31-6	30-10	30-10	30-3	30-3	29-2	24-11	24-2	23-0	22-3
	16.0	30-5	29-11	29-4	29-4	28-9	28-1	27-0	23-0	22-4	21-4	20-7
	19.2	28-8	28-2	27-7	27-1	26-7	25-8	24-8	21-0	20-5	19-6	18-10
	24.0	26-7	26-1[1]	25-7[2]	24-3	23-9	22-11	22-0	18-10	18-3	17-5	16-10

1. The span for No. 1 KD 24 inches o.c. is 2x6 12-5, 2x8 16-5, 2x10 20-11, 2x12 25-5.
2. The span for No. 2 Dense KD 24 inches o.c. is 2x6 12-3, 2x8 16-2, 2x10 20-7, 2x12 25-0.
3. Terms and abbreviations: Sel Str means select structural. KD means KD15, dried to a moisture content of 15 or less, where KD is not shown the material is dried to a moisture content of 19 or less. Lumber dried to 19 or less will be stamped S-DRY or KD S-DRY.
4. These grades may not be commonly available.

TABLE NO. 9. RAFTERS—Low slope (3 in 12 or less)— With no finished ceiling— 30 psf live load.

(Spans are based on a deflection limitation of l/240, except spans in shaded areas are limited by the recommended extreme fiber stress in bending value of the grade and includes a 10 psf dead load.)

Size and Spacing In.	Grade[2] In. o.c.	Dense Sel Str KD and No. 1 Dense KD	Dense Sel Str and Sel Str KD	No. 1 Dense and No. 1 KD	Sel Str	No. 2 Dense KD	No. 1	No. 2 Dense	No. 2 KD	No. 2	No. 3[3] Dense KD	No. 3[3] Dense	No. 3 KD	No. 3
2 x 6	12.0	14-4	14-1	14-1	13-9	13-9	13-9	13-6	13-6	13-2	11-3	10-11	10-5	10-1
	13.7	13-8	13-5	13-5	13-2	13-2	13-2	12-11	12-10	12-4	10-6	10-3	9-9	9-5
	16.0	13-0	12-9	12-9	12-6	12-6	12-6	12-3	11-11	11-5	9-9	9-6	9-0	8-9
	19.2	12-3	12-0	12-0	11-9	11-9	11-6	11-3	10-10	10-5	8-11	8-8	8-3	8-0
	24.0	11-4	11-2	11-1[1]	10-11	10-7	10-3	10-1	9-8	9-4	8-0	7-9	7-4	7-1
2 x 8	12.0	18-10	18-6	18-6	18-2	18-2	18-2	17-10	17-10	17-5	14-10	14-5	13-9	13-3
	13.7	18-0	17-9	17-9	17-5	17-5	17-5	17-0	16-11	16-3	13-11	13-6	12-10	12-5
	16.0	17-2	16-10	16-10	16-6	16-6	16-6	16-2	15-8	15-1	12-10	12-6	11-11	11-6
	19.2	16-1	15-10	15-10	15-6	15-6	15-1	14-10	14-4	13-9	11-9	11-5	10-10	10-6
	24.0	15-0	14-8	14-8[1]	14-5	14-0	13-6	13-3	12-10	12-4	10-6	10-2	9-9	9-5
2 x 10	12.0	24-1	23-8	23-8	23-9	23-2	23-2	22-9	22-9	22-2	18-11	18-5	17-6	16-11
	13.7	23-0	22-7	22-7	22-2	22-2	22-2	21-9	21-7	20-9	17-9	17-2	16-5	15-10
	16.0	21-10	21-6	21-6	21-1	21-1	21-1	20-8	20-0	19-3	16-5	15-11	15-2	14-8
	19.2	20-7	20-2	20-2	19-10	19-10	19-3	18-11	18-3	17-6	15-0	14-6	13-10	13-5
	24.0	19-1	18-9	18-8[1]	18-5	17-10	17-3	16-11	16-4	15-8	13-5	13-0	12-5	12-0
2 x 12	12.0	29-3	28-9	28-9	28-2	28-2	28-2	27-8	27-8	27-0	23-0	22-4	21-4	20-7
	13.7	28-0	27-6	27-6	27-0	27-0	27-0	26-5	26-3	25-3	21-7	20-11	19-11	19-3
	16.0	26-7	26-1	26-1	25-7	25-7	25-7	25-1	24-4	23-4	19-11	19-4	18-6	17-10
	19.2	25-0	24-7	24-7	24-1	24-1	23-5	23-0	22-2	21-4	18-3	17-8	16-10	16-3
	24.0	23-3	22-10	22-8[1]	22-5	21-8	21-0	20-7	19-10	19-1	16-3	15-10	15-1	14-7

1. The span for No. 1 KD, 24 inches o.c. is 2x6, 10-9; 2x8, 14-2; 2x10, 18-1; 2x12, 22-0.
2. Terms and abbreviations: Sel. Str means select structural; KD means KD15, dried to a moisture content of 15% or less. where KD is not shown the material is dried to a moisture content of 19% or less. Lumber dried to 19% or less will be stamped S-DRY or KD S-DRY.
3. These grades may not be commonly available.

TABLE NO. 10. RAFTERS—Low slope (3 in 12 or less)—With no finished ceiling—40 psf live load. (Spans are based on a deflection limitation of l/240, except spans in shaded areas are limited by the recommended extreme fiber stress in bending value of the grade and includes a 10 psf dead load.)

Size and Spacing	Grade [1] In. o.c.	Dense Sel Str KD and No. 1 Dense KD	Dense Sel Str KD	No. 1 Dense	No. 1 KD	Sel Str	No. 2 Dense KD	No. 1	No. 2 Dense	No. 2 KD	No. 2	No. 3 Dense KD [2]	No. 3 Dense [2]	No. 3 KD	No. 3
2 x 6	12.0	13-0	12-9	12-9	12-9	12-6	12-6	12-6	12-3	12-3	11-10	10-1	9-9	9-4	9-0
	13.7	12-5	12-3	12-3	12-3	12-0	12-0	12-0	11-9	11-6	11-0	9-5	9-2	8-9	8-5
	16.0	11-10	11-7	11-7	11-7	11-5	11-5	11-3	11-0	10-8	10-3	8-9	8-6	8-1	7-10
	19.2	11-1	10-11	10-11	10-9	10-8	10-7	10-3	10-1	9-8	9-4	8-0	7-9	7-4	7-1
	24.0	10-4	10-2	9-11	9-8	9-11	9-6	9-6	9-0	8-8	8-4	7-1	6-11	6-7	6-4
2 x 8	12.0	17-2	16-10	16-10	16-10	16-6	16-6	16-6	16-2	16-2	15-7	13-3	12-11	12-4	11-11
	13.7	16-5	16-1	16-1	16-1	15-9	15-9	15-9	15-6	15-2	14-7	12-5	12-1	11-6	11-1·
	16.0	15-7	15-3	15-3	15-3	15-0	15-0	14-10	14-7	14-0	13-6	11-6	11-2	10-8	10-3
	19.2	14-8	14-5	14-5	14-2	14-1	14-0	13-6	13-3	12-10	12-4	10-6	10-2	9-9	9-5
	24.0	13-7	13-4	13-1	12-8	13-1	12-6	12-1	11-11	11-5	11-0	9-5	9-1	8-8	8-5
2 x 10	12.0	21-10	21-6	21-6	21-6	21-1	21-1	21-1	20-8	20-8	19-10	16-11	16-5	15-8	15-2
	13.7	20-11	20-6	20-6	20-6	20-2	20-2	20-2	19-9	19-4	18-7	15-10	15-5	14-8	14-2
	16.0	19-10	19-6	19-6	19-6	19-2	19-2	18-11	18-7	17-11	17-2	14-8	14-3	13-7	13-1
	19.2	18-8	18-4	18-4	18-1	18-0	17-10	17-3	16-11	16-4	15-8	13-5	13-0	12-5	12-0
	24.0	17-4	17-0	16-8	16-2	16-9	15-11	15-5	15-2	14-7	14-0	12-0	11-8	11-1	10-9
2 x 12	12.0	26-7	26-1	26-1	26-1	25-7	25-7	25-7	25-1	25-1	24-2	20-7	20-0	19-1	18-5
	13.7	25-5	25-0	25-0	25-0	24-6	24-6	24-6	24-0	23-6	22-7	19-3	18-9	17-10	17-3
	16.0	24-2	23-9	23-9	23-9	23-3	23-3	23-0	22-7	21-9	20-11	17-10	17-4	16-6	16-0
	19.2	22-9	22-4	22-4	22-0	21-11	21-8	21-0	20-7	19-10	19-1	16-3	15-10	15-1	14-7
	24.0	21-1	20-9	20-4	19-8	20-4	19-5	18-9	18-5	17-9	17-1	14-7	14-2	13-6	13-0

1. Terms and abbreviations: Sel. Str means select structural. KD means KD15, dried to a moisture content of 15% or less; where KD is not shown the material is dried to a moisture content of 19% or less. Lumber dried to 19% or less will be stamped S-DRY or KD S-DRY.

2. These grades may not be commonly available

TABLE NO. 11. RAFTERS—High slope (over 3 in 12)—With no finished ceiling. 20 psf live load + 15 psf dead load—heavy roofing. (Spans are based on a deflection limitation of $l/180$, except spans in shaded areas are limited by the recommended extreme fiber stress in bending value of the grade and includes a 15 psf dead load.)

Size and Spacing in.	in. o.c.	Dense Sel Str KD	Dense Sel Str	No. 1 Dense KD and Sel Str KD	Sel Str	No. 1 Dense	No. 1 KD	No. 2 Dense KD	No. 1	No. 2 Dense	No. 2 KD	No. 2	No. 3 Dense KD[3]	No. 3 Dense[3]	No. 3 KD[3]	No. 3[3]	Construction KD	Construction	Standard KD	Standard
2 x 4	12.0	11-6	11-3	11-6[1]	11-1	11-3	11-2	11-0	10-8	10-6	10-2	9-8	8-2	7-11	7-7	7-3	8-7	8-2	6-6	6-3
	13.7	11-0	10-9	11-0[1]	10-7	10-9	10-5	10-3	10-0	9-10	9-6	9-1	7-8	7-4	7-1	6-9	8-0	7-8	6-1	5-10
	16.0	10-5	10-3	10-5[1]	10-0	10-0	9-8	9-6	9-3	9-1	8-10	8-5	7-1	6-10	6-6	6-3	7-5	7-1	5-7	5-5
	19.2	9-10	9-8	9-6	9-2	9-2	8-10	8-8	8-5	8-4	8-1	7-8	6-6	6-3	6-0	5-8	6-9	6-6	5-1	4-11
	24.0	9-1	8-11	8-6	8-2	8-2	7-11	7-9	7-7	7-5	7-3	6-10	5-9	5-7	5-4	5-1	6-1	5-9	4-7	4-5
2 x 6	12.0	18-0	17-8	17-6	17-0	16-9	16-3	16-0	15-6	15-3	14-8	14-1	12-0	11-8	11-2	10-9				
	13.7	17-3	16-11	16-5	15-11	15-8	15-3	15-0	14-6	14-3	13-9	13-2	11-3	10-11	10-5	10-1				
	16.0	16-4	16-0	15-2	14-9	14-6	14-1	13-11	13-5	13-2	12-9	12-3	10-5	10-1	9-8	9-4				
	19.2	15-1	14-7	13-10	13-6	13-3	12-10	12-8	12-3	12-0	11-7	11-2	9-6	9-3	8-10	8-6				
	24.0	13-6	13-0	12-5	12-0	11-10	11-6	11-4	11-0	10-9	10-5	10-0	8-6	8-3	7-11	7-7				
2 x 8	12.0	23-9	23-4	23-1	22-5	22-1	21-6	21-1	20-5	20-1	19-4	18-7	15-10	15-5	14-8	14-2				
	13.7	22-9	22-4	21-7	21-0	20-8	20-1	19-9	19-1	18-9	18-1	17-5	14-10	14-5	13-9	13-3				
	16.0	21-7	21-0	20-0	19-5	19-2	18-7	18-4	17-8	17-5	16-9	16-1	13-9	13-4	12-9	12-4				
	19.2	19-11	19-2	18-3	17-9	17-6	17-0	16-8	16-2	15-10	15-4	14-8	12-7	12-2	11-7	11-3				
	24.0	17-10	17-2	16-4	15-10	15-8	15-2	14-11	14-5	14-2	13-8	13-2	11-3	10-11	10-5	10-0				
2 x 10	12.0	30-4	29-9	29-5	28-8	28-3	27-5	26-11	26-1	25-7	24-8	23-9	20-3	19-8	18-9	18-1				
	13.7	29-0	28-6	27-7	26-9	26-5	25-7	25-3	24-5	24-0	23-1	22-2	18-11	18-5	17-6	16-11				
	16.0	27-6	26-10	25-6	24-10	24-5	23-9	23-4	22-7	22-2	21-4	20-6	17-6	17-0	16-3	15-8				
	19.2	25-5	24-6	23-3	22-8	22-4	21-8	21-4	20-7	20-3	19-6	18-9	16-0	15-7	14-10	14-4				
	24.0	22-8	21-11	20-10	20-3	19-11	19-4	19-1	18-5	18-1	17-5	16-9	14-4	13-11	13-3	12-10				

1 The span for Select Structural KD 2x4 12 inches o.c. is 11-3. 13 7 inches o.c. 10-9 and 16 inches o.c. 10-3

2 Terms and abbreviations Sel Str means select structural KD means structural KD15 dried to a moisture content of 15 or less where KD is not shown the material is dried to a moisture content of 19½ or less. Lumber dried to 19½ or less will be stamped S-DRY or KD S-DRY

3 These grades may not be commonly available

TABLE NO. 12. RAFTERS—High slope (over 3 in 12)—With no finished ceiling. 30 psf live load + 15 psf dead load—heavy roofing. (Spans are based on a deflection limitation of $l/180$, except spans in shaded areas are limited by the recommended extreme fiber stress in bending value of the grade and includes a 15 psf dead load.)

Size and Spacing in.	in. o.c.	Grade[2] Dense Sel Str KD	Dense Sel Str	No. 1 Dense KD and Sel Str KD	Sel Str	No. 1 Dense	No. 1 KD	No. 2 Dense KD	No. 1	No. 2 Dense	No. 2 KD	No. 2	No. 3 Dense KD[3]	No. 3[3] Dense	No. 3 KD	No. 3	Construction KD	Construction	Standard KD	Standard
2 x 4	12.0	10-0	9-10	10-0[1]	9-8	9-10	9-10	9-8	9-5	9-3	9-0	8-7	7-3	6-11	6-8	6-4	7-7	7-3	5-9	5-6
	13.7	9-7	9-5	9-7[1]	9-3	9-5	9-2	9-1	8-10	8-8	8-5	8-0	6-9	6-6	6-3	5-11	7-1	6-9	5-4	5-1
	16.0	9-1	8-11	9-1[1]	8-9	8-10	8-6	8-5	8-2	8-0	7-9	7-5	6-3	6-0	5-9	5-6	6-7	6-3	4-11	4-9
	19.2	8-7	8-5	8-4	8-1	8-1	7-9	7-8	7-5	7-4	7-1	6-9	5-9	5-6	5-3	5-0	6-0	5-9	4-6	4-4
	24.0	7-11	7-10	7-6	7-3	7-3	6-11	6-10	6-8	6-7	6-4	6-1	5-1	4-11	4-9	4-6	5-4	5-1	4-0	3-10
2 x 6	12.0	15-9	15-9	15-5	15-0	14-10	14-4	14-2	13-8	13-5	12-11	12-5	10-7	10-4	9-10	9-6				
	13.7	15-1	14-9	14-5	14-1	13-10	13-5	13-3	12-9	12-7	12-1	11-8	9-11	9-8	9-2	8-11				
	16.0	14-4	14-1	13-4	13-0	12-10	12-5	12-3	11-10	11-8	11-2	10-9	9-2	8-11	8-6	8-3				
	19.2	13-4	12-10	12-2	11-10	11-8	11-4	11-2	10-10	10-7	10-3	9-10	8-5	8-2	7-9	7-6				
	24.0	11-11	11-6	10-11	10-7	10-6	10-2	10-0	9-8	9-6	9-2	8-10	7-6	7-3	6-11	6-9				
2 x 8	12.0	20-9	20-5	20-4	19-10	19-6	18-11	18-8	18-0	17-8	17-1	16-5	14-0	13-7	12-11	12-6				
	13.7	19-10	19-6	19-0	18-6	18-3	17-8	17-5	16-10	16-7	16-0	15-4	13-1	12-9	12-1	11-9				
	16.0	18-10	18-6	17-7	17-2	16-11	16-5	16-2	15-7	15-4	14-9	14-2	12-1	11-9	11-3	10-10				
	19.2	17-7	16-11	16-1	15-8	15-5	15-0	14-9	14-3	14-0	13-6	12-11	11-1	10-9	10-3	9-11				
	24.0	15-8	15-2	14-5	14-0	13-10	13-5	13-2	12-9	12-6	12-1	11-7	9-11	9-7	9-2	8-10				
2 x 10	12.0	26-6	26-0	26-0	25-3	24-11	24-2	23-9	23-0	22-7	21-9	20-11	17-10	17-4	16-6	16-0				
	13.7	25-4	24-11	24-4	23-8	23-3	22-7	22-3	21-6	21-2	20-4	19-7	16-8	16-3	15-6	14-11				
	16.0	24-1	23-8	22-6	21-10	21-7	20-11	20-7	19-11	19-7	18-10	18-1	15-6	15-0	14-4	13-10				
	19.2	22-5	21-7	20-6	20-0	19-8	19-1	18-9	18-2	17-10	17-2	16-6	14-1	13-8	13-1	12-8				
	24.0	20-0	19-4	18-4	17-10	17-7	17-1	16-10	16-3	16-0	15-5	14-9	12-8	12-3	11-8	11-4				

1. The span for Select Structural KD 2x4 12 inches o.c. is 9-10, 13.7 inches o.c. 9-5, and 16 inches o.c. 8-11
2. Terms and abbreviations. Sel Str means select structural. KD means dried to a moisture content of 15% or less, where KD is not shown the material is dried to a moisture content of 19% or less. Lumber dried to 19% or less will be stamped S-DRY or KD S-DRY
3. These grades may not be commonly available

TABLE NO. 13. RAFTERS—High slope (over 3 in 12)—With no finished ceiling—20 psf live load + 7 psf dead load—light roofing.

(Spans are based on a deflection limitation $l/180$, except spans in shaded areas are limited by the recommended extreme fiber stress in bending value of the grade and includes a 7 psf dead load.)

Size in.	Spacing in. o.c.	Dense Sel Str KD	Sel Str KD and Dense Sel Str	No. 1 Dense KD	Sel Str	No. 1 Dense	No. 1 KD	No. 2 Dense KD	No. 1	No. 2 Dense	No. 2 KD	No. 2	No. 3[2] Dense KD	No. 3 Dense	No. 3 KD	No. 3	Construction KD	Construction	Standard KD	Standard
2 x 4	12.0	11-6	11-3	11-6	11-1	11-3	11-3	11-1	11-1	10-10	10-10	10-10	9-4	9-0	8-7	8-3	9-9	9-4	7-4	7-1
	13.7	11-0	10-9	11-0	10-7	10-9	10-9	10-7	10-7	10-4	10-4	10-4	8-9	8-5	8-1	7-8	9-2	8-9	6-11	6-7
	16.0	10-5	10-3	10-5	10-0	10-3	10-3	10-0	10-0	9-10	9-10	9-7	8-1	7-9	7-5	7-1	8-6	8-1	6-5	6-1
	19.2	9-10	9-8	9-10	9-5	9-8	9-8	9-5	9-5	9-3	9-2	8-9	7-4	7-1	6-10	6-6	7-9	7-4	5-10	5-7
	24.0	9-1	8-11	9-1	8-9	8-11	8-11	8-7	8-7	8-6	8-3	7-10	6-7	6-4	6-1	5-10	6-11	6-7	5-3	5-0
2 x 6	12.0	18-0	17-8	18-0	17-4	17-8	17-8	17-4	17-4	17-0	16-8	16-1	13-8	13-4	12-8	12-3				
	13.7	17-3	16-11	17-3	16-7	16-11	16-11	16-7	16-6	16-3	15-8	15-0	12-10	12-5	11-10	11-6				
	16.0	16-4	16-1	16-4	15-9	16-1	16-1	15-9	15-3	15-0	14-6	13-11	11-10	11-6	11-0	10-7				
	19.2	15-5	15-2	15-5	14-10	15-1	14-8	14-5	13-11	13-8	13-3	12-8	10-10	10-6	10-0	9-8				
	24.0	14-4	14-1	14-1	14-1	13-6	13-1	12-11	12-6	12-3	11-10	11-4	9-8	9-5	9-0	8-8				
2 x 8	12.0	23-9	23-4	23-9	22-11	23-4	23-4	22-11	22-11	22-5	22-0	21-2	18-1	17-7	16-9	16-2				
	13.7	22-9	22-4	22-9	21-11	22-4	22-4	21-11	21-9	21-5	20-7	19-10	16-11	16-5	15-8	15-1				
	16.0	21-7	21-2	21-7	20-10	21-2	21-2	20-10	20-2	19-10	19-1	18-4	15-8	15-2	14-6	14-0				
	19.2	20-4	19-11	20-4	19-7	19-11	19-4	19-0	18-5	18-1	17-5	16-9	14-3	13-10	13-3	12-9				
	24.0	18-10	18-6	18-7	18-1	17-10	17-3	17-0	16-5	16-2	15-7	15-0	12-9	12-5	11-10	11-5				
2 x 10	12.0	30-4	29-9	30-4	29-2	29-9	29-9	29-2	29-2	28-7	28-1	27-0	23-1	22-5	21-4	20-7				
	13.7	29-0	28-6	29-0	27-11	28-6	28-6	27-11	27-9	27-3	26-3	25-3	21-7	20-11	20-0	19-3				
	16.0	27-6	27-1	27-6	26-6	27-1	27-0	26-6	25-8	25-3	24-4	23-5	20-0	19-5	18-6	17-10				
	19.2	25-11	25-5	25-11	25-0	25-5	24-8	24-3	23-6	23-1	22-3	21-4	18-3	17-8	16-10	16-4				
	24.0	24-1	23-8	23-8	23-1	22-9	22-1	21-8	21-0	20-7	19-10	19-1	16-4	15-10	15-1	14-7				

1. Terms and abbreviations. Sel. Str. means select structural. KD means KD15, dried to a moisture content of 15% or less, where KD is not shown the material is dried to a moisture content of 19% or less. Lumber dried to 19% or less will be stamped S-DRY or KD S-DRY

2. These grades may not be commonly available

TABLE NO. 14. RAFTERS—High slope (over 3 in 12)—With no finished ceiling—30 psf live load + 7 psf dead load—light roofing. (Spans are based on a deflection limitation of $l/180$, except spans in shaded areas are limited by the recommended extreme fiber stress in bending value of the grade and includes a 7 psf dead load.)

Size	Spacing in.o.c.	Dense Sel Str KD	Sel Str KD and Dense Sel Str	No. 1 Dense KD	Sel Str	No. 1 Dense	No. 1 KD	No. 2 Dense KD	No. 1	No. 2 Dense	No. 2 KD	No. 2	No. 3 Dense KD[3]	No. 3 Dense[3]	No. 3 KD[3]	No. 3[3]	Construction KD	Construction	Standard KD	Standard
2 x 4	12.0	10-0	9-10	10-0	9-8	9-10	9-10	9-8	9-8	9-6	9-6	9-5	8-0	7-8	7-4	7-0	8-4	8-0	6-4	6-0
	13.7	9-7	9-5	9-7	9-3	9-5	9-5	9-3	9-3	9-1	9-1	8-10	7-5	7-2	6-10	6-7	7-10	7-5	5-11	5-8
	16.0	9-1	8-11	9-1	8-9	8-11	8-11	8-9	8-9	8-7	8-7	8-2	6-11	6-8	6-4	6-1	7-3	6-11	5-5	5-3
	19.2	8-7	8-5	8-7	8-3	8-5	8-5	8-3	8-3	8-1	7-10	7-5	6-4	6-1	5-10	5-7	6-7	6-4	5-0	4-9
	24.0	7-11	7-10	7-11	7-8	7-10	7-8	7-7	7-4	7-3	7-0	6-8	5-8	5-5	5-2	5-0	5-11	5-8	4-5	4-3
2 x 6	12.0	15-9	15-6	15-9	15-2	15-6	15-6	15-2	15-1	14-10	14-3	13-9	11-9	11-4	10-10	10-6				
	13.7	15-1	14-9	15-1	14-6	14-9	14-9	14-6	14-1	13-10	13-4	12-10	10-11	10-8	10-2	9-10				
	16.0	14-4	14-1	14-4	13-9	14-1	13-9	13-6	13-1	12-10	12-4	11-11	10-2	9-10	9-5	9-1				
	19.2	13-6	13-3	13-6	13-0	12-11	12-6	12-4	11-11	11-9	11-3	10-10	9-3	9-0	8-7	8-3				
	24.0	12-6	12-3[1]	12-0	11-9	11-6	11-2	11-0	10-8	10-6	10-1	9-8	8-3	8-0	7-8	7-5				
2 x 8	12.0	20-9	20-5	20-9	20-0	20-5	20-5	20-0	19-10	19-6	18-10	18-1	15-5	15-0	14-3	13-10				
	13.7	19-6	19-6	19-10	19-2	19-6	19-6	19-2	18-7	18-3	17-7	16-11	14-5	14-0	13-4	12-11				
	16.0	18-6	18-6	18-10	18-2	18-6	18-1	17-9	17-2	16-11	16-4	15-8	13-4	13-0	12-5	11-11				
	19.2	17-9	17-5	17-9	17-1	17-0	16-6	16-3	15-9	15-5	14-10	14-3	12-2	11-10	11-4	10-11				
	24.0	16-6	16-2[1]	15-10	15-5	15-3	14-9	14-6	14-1	13-10	13-4	12-9	10-11	10-7	10-1	9-9				
2 x 10	12.0	26-6	26-0	26-6	25-6	26-0	26-0	25-6	25-4	24-11	24-0	23-1	19-8	19-1	18-3	17-7				
	13.7	25-4	24-11	25-4	24-5	24-11	24-11	24-5	23-9	23-4	22-5	21-7	18-5	17-11	17-1	16-6				
	16.0	24-1	23-8	24-1	23-2	23-8	23-1	22-8	21-11	21-7	20-9	20-0	17-1	16-7	15-9	15-3				
	19.2	22-8	22-3	22-8	21-10	21-8	21-1	20-9	20-1	19-8	19-0	18-3	15-7	15-1	14-5	13-11				
	24.0	21-0	20-8[1]	20-3	19-8	19-5	18-10	18-6	17-11	17-7	17-0	16-4	13-11	13-6	12-11	12-5				

1. The span for Select Structural KD 24 inches o.c. is 2x6, 12-0; 2x8, 15-10; 2x10, 20-3.
2. Terms and abbreviations. Sel Str means select structural. KD means KD15, dried to a moisture content of 15% or less, where KD is not shown the material is dried to a moisture content of 19% or less. Lumber dried to 19% or less will be stamped S-DRY or KD S-DRY.
3. These grades may not be commonly available.

Selected References

Maguire, Byron W. *Complete Book of Furniture and Cabinetmaking*, Reston Publishing, Reston, VA.

Drywall Construction Handbook, United States Gypsum, Chicago, IL.

Exteriors, Hardboard Siding, Masonite Corporation, Chicago, IL.

General Information, Lumber, Western Wood Products Association, Portland, OR.

Hinged Interior Wood Door Units (PS-32-70), United States Department of Commerce, Washington, D.C.

MOD 24 Building Guide, American Plywood Association, Tacoma, Washington.

Plywood Commercial/Industrial Construction Guide, American Plywood Association, Tacoma, Washington.

Plywood for Concrete Forming, American Plywood Association, Tacoma, Washington.

Plywood Residential Construction Guide, American Plywood Association, Tacoma, Washington.

Recommended Practice for Concrete Formwork (AC1 347-68), American Concrete Institute, Detroit, MI.

Suspended Ceilings, Armstrong World Industries, Lancaster, PA.

Southern Pine Maximum Spans for Joists and Rafters, Southern Forest Products Association, New Orleans, LA.

Uniform Building Code, Whittier, CA.

Working Stresses for Joists and Rafters, Western Wood Products Association, Portland, OR.

Index

Other Practical References

Carpentry Estimating

Simple, clear instructions show you how to take off quantities and figure costs for all rough and finish carpentry. Shows how much overhead and profit to include, how to convert piece prices to MBF prices or linear foot prices, and how to use the tables included to quickly estimate manhours. All carpentry is covered: floor joists, exterior and interior walls and finishes, ceiling joists and rafters, stairs, trim, windows, doors, and much more. Includes sample forms, checklists, and the author's factor worksheets to save you time and help prevent errors. **320 pages, 8½ x 11, $25.50**

Roof Framing

Frame any type of roof in common use today, even if you've never framed a roof before. Shows how to use a pocket calculator to figure any common, hip, valley, and jack rafter length in seconds. Over 400 illustrations take you through every measurement and every cut on each type of roof: gable, hip, Dutch, Tudor, gambrel, shed, gazebo and more. **480 pages, 5½ x 8½, $19.50**

Contractor's Guide to the Building Code

Explains in plain English exactly what the Uniform Building Code requires and shows how to design and construct residential and light commercial buildings that will pass inspection the first time. Suggests how to work with the inspector to minimize construction costs, what common building short cuts are likely to be cited, and where exceptions are granted. **312 pages, 5½ x 8½, $16.25**

Rough Carpentry

All rough carpentry is covered in detail: sills, girders, columns, joists, sheathing, ceiling, roof and wall framing, roof trusses, dormers, bay windows, furring and grounds, stairs and insulation. Many of the 24 chapters explain practical code approved methods for saving lumber and time without sacrificing quality. Chapters on columns, headers, rafters, joists and girders show how to use simple engineering principles to select the right lumber dimension for whatever species and grade you are using. **288 pages, 8½ x 11, $16.00**

Finish Carpentry

The time-saving methods and proven shortcuts you need to do first class finish work on any job: cornices and rakes, gutters and downspouts, wood shingle roofing, asphalt, asbestos and built-up roofing, prefabricated windows, door bucks and frames, door trim, siding, wallboard, lath and plaster, stairs and railings, cabinets, joinery, and wood flooring. **192 pages, 8½ x 11, $10.50**

Building Cost Manual

Square foot costs for residential, commercial, industrial, and farm buildings. In a few minutes you work up a reliable budget estimate based on the actual materials and design features, area, shape, wall height, number of floors and support requirements. Most important, you include all the important variables that can make any building unique from a cost standpoint. **240 pages, 8½ x 11, $14.00. Revised annually**

National Construction Estimator

Current building costs in dollars and cents for residential, commercial and industrial construction. Prices for every commonly used building material, and the proper labor cost associated with installation of the material. Everything figured out to give you the "in place" cost in seconds. Many time-saving rules of thumb, waste and coverage factors and estimating tables are included. **528 pages, 8½ x 11, $18.50. Revised annually.**

Rafter Length Manual

Complete rafter length tables and the "how to" of roof framing. Shows how to use the tables to find the actual length of common, hip, valley and jack rafters. Shows how to measure, mark, cut and erect the rafters, find the drop of the hip, shorten jack rafters, mark the ridge and much more. Has the tables, explanations and illustrations every professional roof framer needs. **369 pages, 8½ x 5½, $12.25**

Handbook of Construction Contracting Vol. 1 & 2

Volume 1: Everything you need to know to start and run your construction business; the pros and cons of each type of contracting, the records you'll need to keep, and how to read and understand house plans and specs to find any problems before the actual work begins. All aspects of construction are covered in detail, including all-weather wood foundations, practical math for the jobsite, and elementary surveying. **416 pages, 8½ x 11, $21.75**

Volume 2: Everything you need to know to keep your construction business profitable; different methods of estimating, keeping and controlling costs, estimating excavation, concrete, masonry, rough carpentry, roof covering, insulation, doors and windows, exterior finish, specialty finishes, scheduling work flow, managing workers, advertising and sales, spec building and land development and selecting the best legal structure for your business. **320 pages, 8½ x 11, $24.75**

Easy-To-Use 10 Day Examination Cards

Mail This No Risk Card Today

☐ Please send me the books checked for 10 days free examination. At the end of that time I will pay in full plus postage (& 6% tax in Calif.) or return the books postpaid and owe nothing.
☐ Enclosed is my full payment or Visa/MasterCard number. Please rush me the books without charging for postage.

☐ 14.00 Building Cost Manual
☐ 25.50 Carpentry Estimating
☐ 16.25 Contractor's Guide to the Building Code
☐ 10.50 Finish Carpentry
☐ 21.75 Handbook of Construction Contracting Vol. 1
☐ 24.75 Handbook of Construction Contracting Vol. 2
☐ 18.50 National Construction Estimator
☐ 12.25 Rafter Length Manual
☐ 19.50 Roof Framing
☐ 16.00 Rough Carpentry

In a Hurry?
We accept phone orders charged to your MasterCard or Visa.
Call (619) 438-7828

Craftsman

Name (Please print clearly)

Company

Address

City/State/Zip

☐ Visa ☐ Master Card

Exp. date_____

Number

CICC card

Mail This No Risk Card Today

☐ Please send me the books checked for 10 days free examination. At the end of that time I will pay in full plus postage (& 6% tax in Calif.) or return the books postpaid and owe nothing.
☐ Enclosed is my full payment or Visa/MasterCard number. Please rush me the books without charging for postage.

☐ 14.00 Building Cost Manual
☐ 25.50 Carpentry Estimating
☐ 16.25 Contractor's Guide to the Building Code
☐ 10.50 Finish Carpentry
☐ 21.75 Handbook of Construction Contracting Vol. 1
☐ 24.75 Handbook of Construction Contracting Vol. 2
☐ 18.50 National Construction Estimator
☐ 12.25 Rafter Length Manual
☐ 19.50 Roof Framing
☐ 16.00 Rough Carpentry

In a Hurry?
We accept phone orders charged to your MasterCard or Visa.
Call (619) 438-7828

Craftsman

Name (Please print clearly)

Company

Address

City/State/Zip

☐ Visa ☐ Master Card

Exp. date_____

Number

CICC card

In a hurry?
We accept charge card phone orders.
Call (619) 438-7828

NO POSTAGE
NECESSARY
IF MAILED
IN THE
UNITED STATES

BUSINESS REPLY MAIL

FIRST CLASS MAIL   PERMIT NO.271   CARLSBAD, CA

POSTAGE WILL BE PAID BY ADDRESSEE

Craftsman Book Company
6058 Corte Del Cedro
P. O. Box 6500
Carlsbad, CA 92008—9974

NO POSTAGE
NECESSARY
IF MAILED
IN THE
UNITED STATES

BUSINESS REPLY MAIL

FIRST CLASS MAIL   PERMIT NO.271   CARLSBAD, CA

POSTAGE WILL BE PAID BY ADDRESSEE

Craftsman Book Company
6058 Corte Del Cedro
P. O. Box 6500
Carlsbad, CA 92008—9974